Analog Circuits and Signal Processing

More information about this series at http://www.springer.com/series/7381

Yan Lu • Wing-Hung Ki

CMOS Integrated Circuit Design for Wireless Power Transfer

Springer

Yan Lu
University of Macau
Macao, China

Wing-Hung Ki
The Hong Kong University of Science and Technology
Hong Kong, China

ISSN 1872-082X ISSN 2197-1854 (electronic)
Analog Circuits and Signal Processing
ISBN 978-981-10-2614-0 ISBN 978-981-10-2615-7 (eBook)
DOI 10.1007/978-981-10-2615-7

Library of Congress Control Number: 2017945086

Printed on acid-free paper

This Springer imprint is published by Springer Nature
The registered company is Springer Nature Singapore Pte Ltd.
The registered company address is: 152 Beach Road, #21-01/04 Gateway East, Singapore 189721, Singapore

Preface

CMOS integrated circuit design for wireless power transfer intends to report the state-of-the-art analog and power management IC design techniques for various wireless power transfer (WPT) systems. To propose elaborate power management solutions, the circuit designers are required to have in-depth understanding on the characteristics of each type of converters and regulators in the power chain. This book addresses the design issues of WPT at both the system level and the circuit level and serves as a handbook with design hints for research students and engineers in analog integrated circuit design, integrated power electronics, and wireless power transfer system design.

Our research has been focusing on fully integrated power management integrated circuits design that aims at miniaturizing the devices by proposing novel circuit topologies and system architectures. Since the passive components (inductors and capacitors) occupy most of the chip and board area, reducing their values and numbers and reusing them are the keys to minimizing the volume of the portable/wearable/implantable devices. Reducing the supply noise also provides more design margin for the functional loading circuits, and increasing the power conversion efficiency can maximize the available energy and prolong the device operation time on a single charge of the battery.

Besides discussing the power management solutions for biomedical implantable systems, we also touch upon the topics of wireless charging for portable and wearable devices, wireless powering for storage devices, and RF energy harvesting for internet-of-things (IoT). New regulation methods and state-of-the-art converter architectures at both the system level and the circuit level are briefed. Our work focuses on the CMOS integrated circuit design of wireless power transfer systems and active circuits that effectively convert the power between the AC and DC domains. Voltage regulation at both the system level and the circuit level would also be addressed. We hope our work could help electronic engineers and students in designing high-quality power supplies for the wirelessly powered devices.

Macao, China Yan Lu
Hong Kong, China Wing-Hung Ki
April 2017

Contents

1 Introduction of Wireless Power Transfer 1
1.1 Motivations 1
1.2 Operation Principles, Regulations, and Standards 2
1.2.1 Near-Field and Far-Field Operation 3
1.2.2 Ultrasound Wireless Power Transfer 5
1.2.3 Inductive and Resonant WPT 6
1.2.4 Wireless Charging Standards 6
1.3 Design Perspectives 7
1.3.1 Comparison of Wired and Wireless Power 7
1.3.2 System Overview 8
1.4 Organization of the Book 8
References 10

2 Wireless Power Transfer Systems 13
2.1 Introduction 13
2.2 Output Voltage Regulation Schemes 15
2.2.1 Primary Side Non-linear Power Control 16
2.2.2 Reconfigurable Rectifier for Adaptive Output 19
2.2.3 Resonant Regulating Rectifier 21
2.2.4 Reconfigurable Resonant Regulating Rectifiers 23
2.2.5 Pre-rectifier Regulation 25
2.2.6 Multi-Level Single-Inductor Multiple Output Operation 27
2.3 Summary and Discussion 29
References 30

3 Analysis of Coupled-Coils 33
3.1 Introduction 33
3.2 Coupled-Coils and Modeling 34
3.2.1 Ideal Transformer 34
3.2.2 Ideal Coupled-Coils 34

3.2.3 T-Model of Coupled-Coils 35
3.2.4 Transformer Model of Coupled-Coils 36
3.2.5 Reflected Impedance Model of Coupled-Coils 37
3.2.6 Link Voltage Gain and Link Efficiency 38
3.2.7 Computing A and η of the Ideal Coupled-Coils 39
3.2.8 Design-Oriented Analysis of Coupled-Coils with Parasitics 41
3.3 Resonant Coupled-Coils 44
3.3.1 Series-Series Resonant Coupled-Coils 44
3.3.2 Series-Parallel Resonant Coupled-Coils 47
3.4 Summary 50
References 51

4 Circuit Design of CMOS Rectifiers 53
4.1 Introduction 53
4.2 Diodes and Diode-Connected Transistors 55
4.3 Comparator-Based Active Rectifiers 59
4.3.1 Delay Compensation Schemes 61
4.3.2 Delay Time Analysis of Active Diodes 67
4.3.3 Biasing Circuits of Active Rectifiers 70
4.3.4 Full-Wave Rectifier Design Examples 72
4.3.5 Reconfigurable Rectifier Design Example 83
4.4 DLL-Based Rectifiers 86
4.5 Rectifiers for RF Energy Harvesting 89
4.6 Summary and Discussion 92
References 94

5 Linear Regulators for WPT 97
5.1 Introduction 97
5.2 PMOS and NMOS LDO Regulators 99
5.3 Control Loop Design 101
5.3.1 Dominant Pole Considerations 101
5.3.2 Replica Regulator 103
5.3.3 Flipped Voltage Follower 103
5.3.4 Impedance Attenuation Buffer 105
5.3.5 Digitally Controlled LDO Regulator 105
5.4 Design Case Study 106
5.4.1 Design of Tri-Loop LDO Regulator 106
5.4.2 LDO Regulator with Enhanced Super Source Follower 120
5.5 Summary and Discussion 123
References 125

6 DC-DC Converters for WPT 127
6.1 Introduction 127
6.2 DC-DC Converter Comparisons 128

6.3 Control Loop Design 130
6.3.1 Loop Design for Inductor-Based Converters 130
6.3.2 Loop Design for SC DC-DC Converters 132
6.4 Architectures of DC-DC Conversion 136
6.4.1 Multiple-Output Converters 136
6.4.2 Layout Strategy for Efficient Power Delivery 136
6.4.3 Cascade Voltage Regulators 138
6.5 Summary 139
References 140

7 Power Amplifiers for WPT 143
7.1 Introduction 143
7.1.1 Integration Processes 144
7.1.2 Losses in Switching PAs 144
7.1.3 Zero Voltage (Current) Switching 147
7.2 Class-D PA 147
7.2.1 Operation Conditions 149
7.2.2 ZVS Class-D PA 152
7.3 Class-E PA 153
7.4 Summary and Discussion 155
References 156

8 Conclusions and Future Works 159
8.1 Concluding Remarks of the Book 159
8.2 Suggested Future Works 160
References 161

Chapter 1
Introduction of Wireless Power Transfer

Abstract Wireless power transfer (WPT) for a broad range of applications is projected to have an exponential growth, with an enormous number of new devices and products to be enabled by this powerful technology. In this chapter, the background and motivations of WPT are introduced first. Then high-level considerations on WPT such as operation frequencies, WPT regulations and WPT standards are reviewed and summarized. In addition, design perspectives on the WPT circuits and systems are also examined. Finally, we present the organization of this book and provide some reading guidelines.

Keywords Wireless power transfer • Inductive coupling • Wireless charging • Near field • Far field • Ultrasound • Optimum coupling distance

1.1 Motivations

Wireless power transfer (WPT) has a wide range of applications including (arranged from low to high power levels) radio frequency identification (RFID), internet-of-things (IoT), implantable medical devices (IMDs), real-time wireless power for non-contact memory devices and wafer-level testing, and also wireless chargers for portable/wearable devices and electric vehicles (EVs). It is evident that the utilization of WPT technologies is on the verge of exponential growth.

Design considerations for WPT at different power levels are quite different, and are discussed as follows. In the extreme low-power cases, IMDs [1–3] such as the pacemaker, cochlear implant, retinal prosthesis, neural recording microsystem, etc. only consumes a power level of milli-Watt or even lower. The form factor of the IMD power receiver is one of the most important concerns because it has to be as non-invasive as possible. For IMDs with a single channel or a few channels such as the pacemaker and the cochlear implant, a power level of micro-Watt would be sufficient [1]. For retinal prosthesis and neural signal recording, on the other hand, hundreds or thousands of channels are needed [2, 3], as simulations suggest that, for example, 600–1000 pixels will be required to provide visual function such as face recognition and reading [2]. In such a case, the power level would be in the milli-Watt range, and fully on-chip integration is preferred such that no discrete component is needed, and the complete system could occupy just a few cubic millimeters.

Y. Lu, W.-H. Ki, *CMOS Integrated Circuit Design for Wireless Power Transfer*, Analog Circuits and Signal Processing, DOI 10.1007/978-981-10-2615-7_1

High efficiency is of utmost importance, such that only little power is lost and being absorbed by the tissue that would cause potential damage.

In the high-power regime such as charging up automotive battery systems for EVs, discrete high-voltage diodes and transistors are used to handle the power in the Kilo-Watt range. In between IMD powering and EV charging, wireless charging for portable/wearable devices is in the range of a few Watts. For consumer electronics, small size and compact packaging are preferred. The power level of a typical wireless mobile phone charger is around 5 W, and the rectifier is preferably integrated on-chip to avoid using discrete Schottky diodes or III-V transistors and to reduce the cost.

As mentioned above, one potential application of WPT is the contact-less memory [4–7]. There are a few advantages to equip memory cards with wireless data and wireless power transfer functions. First, by using WPT to power up the card, the metal pads in conventional memory cards can be removed, and data transmission rate can be enhanced, as the electrostatic discharge (ESD) protection circuits that introduce additional capacitive loads to the I/O can be removed. Second, for a contact-less memory card, there will be no wear and tear of the card due to repeated plugging and unplugging. Third, the fully sealed packaging enables the device to be waterproof, and more importantly, it dramatically extends the memory lifetime by isolating the chip from moisture and oxygen [4].

Therefore, wireless power transfer is not just about getting rid of wires. It helps many technologies and applications to become possible. For example, unmanned battery-powered aircrafts (drones) can be recharged in mid-air to extend their limited navigation distance; and portable devices can have waterproof packaging with no electrical connectors.

1.2 Operation Principles, Regulations, and Standards

One major concern on wireless power transfer is the health issue. Radiation can be classified into ionizing and non-ionizing radiation, depending on the energy of the radiating particles. Ionizing radiation has energy high enough to ionize atoms and molecules and break chemical bonds; while non-ionizing radiation only generates heat. This is an important distinction that differentiates if the radiation is harmful to living subjects. Ionizing radiations such as Gamma rays (from radioactive decay of atomic nuclei), X-rays (used in medical imaging and security check) and the higher energy range of ultraviolet light (exists in the sunlight) constitute the ionizing part of the electromagnetic spectrum. While visible light, infrared light, microwaves and radio waves have lower energy and longer wavelength, and are non-ionizing. Research showed that the adverse effects due to RF exposure are basically related to the thermal aspects [8].

As summarized in [8], an extensive literature review on RF biological effects that consists of over a thousand primary peer-reviewed publications reveals no adverse health effects that are not thermally related. Behavioral studies on animal subjects indicate that disruption of behavior is usually (not always)

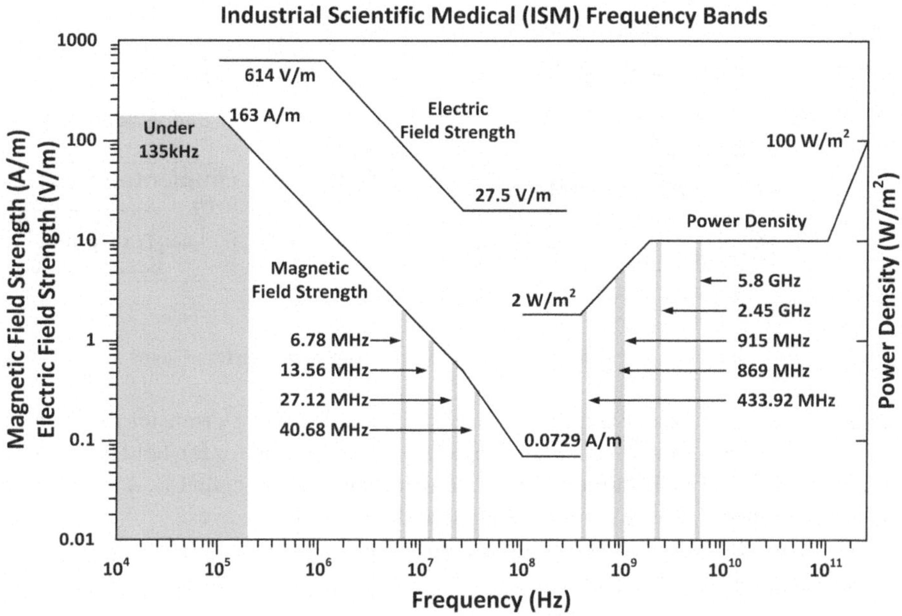

Fig. 1.1 Maximum permissible exposure for the general public versus the ISM frequency bands

accompanied by a core body temperature increase of ~1.0 °C; and a body temperature increase of 2–2.5 °C will cause significant heat-induced abnormalities, which mostly occur after RF exposures for tens of minutes or one hour. Therefore, basic restrictions (BRs) on the whole-body-average for frequencies between 100 kHz and 3 GHz is 0.4 W/kg with a traditional safety factor of ten, and the localized exposure BRs is 4 W/kg averaged over any 10 g of tissue of which the volume is approximately 10 cm^3. The maximum permissible exposure (MPE) versus the ISM (industrial, scientific, medical) frequency bands are shown in Fig. 1.1. Measurement results show that the human tissue specific absorption rate (SAR) increases as the frequency increases [9]. Meanwhile, pulsed waveforms with low duty-cycle can be employed for EM wave transmission to reduce the heat produced, because the heating capability of electric current is proportional to the square root of the duty-cycle [8].

1.2.1 Near-Field and Far-Field Operation

Electromagnetic (EM) WPT systems can basically be divided into two categories according to the transmission modes of near-field and far-field operations. As shown in Fig. 1.2a, near-field operation assumes that the transmission distance d is much shorter than the wavelength λ, that is, $d << \lambda$. In the near-field case, both

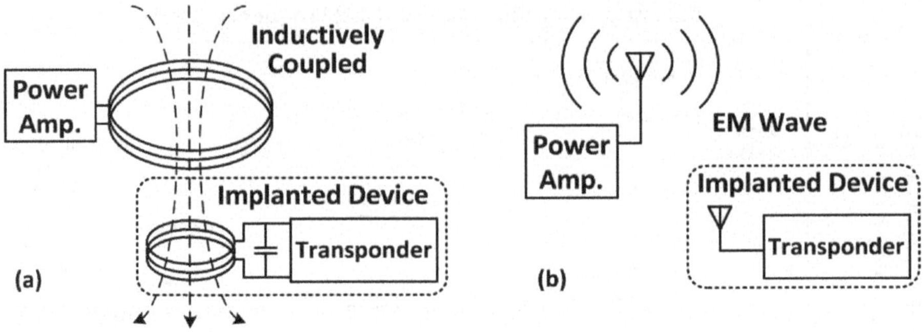

Fig. 1.2 Block diagrams of near-field and far-field wireless power transfer systems

the transmitting and receiving devices are coils, and the energy transfer medium is magnetic flux only. On the other hand, as shown in Fig. 1.2b, for far-field operation, the wavelength is much shorter than the transmission distance, that is, $\lambda << d$, and the energy transfer devices are antennas that deal with EM waves.

For near-field operation, the optimum transmission distance x_{OPT} is related to the radius R of the transmitting coil. A rule-of-thumb optimum distance is suggested in [10] as

$$x_{OPT} \approx R/\sqrt{2}, \tag{1.1}$$

As shown in the conceptual diagram of Fig. 1.3, if the transmitting coil is too large, the field strength becomes very weak, even at a distance of $x = 0$. On the other hand, if the coil is too small, the magnetic field strength drops at a fast rate that is inversely proportional to x^3. Therefore, the size of the transmitting coil should be designed according to the targeted distance.

For far-field operation, it is quite difficult to focus the transmitted energy to a targeted location. Hence, antenna array technologies may be used for direction control of the transmitted power and consequently achieve higher receiver and system efficiencies [11, 12]. Equation (1.2) shows the simplified formula for the path loss A_{PATH} in free space between two isotropic transmitting and receiving antennas:

$$A_{PATH} = -20\log\left(\frac{4\pi d}{\lambda}\right), \tag{1.2}$$

The path loss is inversely proportional to the wavelength, and therefore it is proportional to the frequency. To reduce path loss, lower frequency bands are conventionally used for long distance RF transmission. However, there is another aspect of consideration. By increasing the transmission frequency by ten times, the values of both the inductor and the capacitor can be reduced by ten times, because the resonant frequency F_{RES} is given by

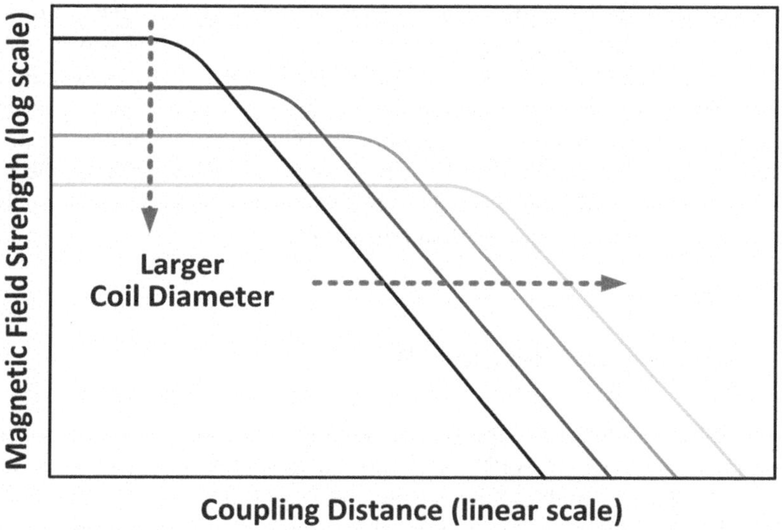

Fig. 1.3 Conceptual diagram of the relationship between the coil diameter and the optimum coupling distance

$$F_{RES} = \frac{1}{2\pi\sqrt{LC}} \tag{1.3}$$

Hence, high-frequency wireless power transfer could reduce the volume of the LC resonant tank of near-field operation or the volume of the antenna of far-field operation. The value of the output filtering capacitor can also be reduced, such that the entire WPT receiver could be integrated on-chip or on-package. Besides, smaller inductance with higher Q can be more easily achieved at higher frequencies within a limited space, which is one of the key factors in improving the overall efficiency of the wireless power transfer systems. For near-field operation, the readily available ISM frequency 13.56 MHz is commonly used for implantable medical devices (IMDs) [13–15]. Although the SAR of tissue increases with frequency in the range of tens to hundreds of MHz, the human tissue SAR for 13.56 MHz is still quite low compared to the thermal power dissipation in the coils [16]. Alternatively, when larger wireless power levels are required, frequencies such as 100–200 kHz and 6.78 MHz are widely adopted for wireless charging [17–19].

1.2.2 Ultrasound Wireless Power Transfer

Ultrasound with a frequency of 20 kHz or above is also a good candidate for wireless power transfer. The advantage of acoustic waves is that the propagation loss in water is much lower than that of electromagnetic (EM) waves [20]. Moreover, the relatively low operating frequency compared to the EM waves makes it easier to achieve higher power conversion efficiency for the power converters used. For safe wireless power transmission in tissue, ultrasound of high power density is allowed, for example, an ultrasonic intensity of 100 mW/cm^2 would generate little or no hazard for the duration of at least two hours [21]. Nevertheless, the power level of ultrasound should still be limited, as it causes much distress to animals that are sensitive to higher frequencies than humans [22].

1.2.3 Inductive and Resonant WPT

It is easy to be confused with the two marketing terms: inductive power transfer (IPT) and resonant WPT (R-WPT). Aren't all the EM WPT systems both inductive and resonant? (This is similar to the question: isn't all food organic?) Yes, both IPT and R-WPT utilize the magnetic field and the resonant circuits. The difference is in the operating mechanism of the WPT transmitter.

A wireless power transfer system can be considered as tightly or loosely coupled by the values of the coupling coefficient k and the distance d of the coils. When the two coils are of similar sizes and are placed very closely together, they are tightly coupled. A higher k improves the WPT efficiency, and consequently reduces the losses and heat. In a tightly coupled system, the transmission frequency is designed to be slightly different from the resonant frequency of the receiver for output power control, and a high efficiency can still be maintained [23]. For a loosely coupled system, the transmission frequency is well-tuned at the receiver resonant frequency to induce sufficient power and to achieve a better efficiency at extended distance, for example, a few times of the diameter of the transmitter coil. The secondary coil is designed to have a very high quality factor Q, and efficient power transfer can be achieved if the transmission frequency is exactly the same as the resonant frequency of the receiver. However, resonant excitation is sensitive to component variations, and hence additional resonant tuning circuits may have to be implemented.

1.2.4 Wireless Charging Standards

Wireless charging usually makes use of the frequency bands in the range of 100 kHz to 10 MHz. There are currently two main standards in the market. One is the Qi (pronounced "Chee", derived from the Chinese word meaning "life energy") standard developed by the Wireless Power Consortium (WPC). The other one is the AirFuel Alliance which is formed from merging the Alliance for Wireless Power (A4WP) and the Power Matters Alliance (PMA). Both standards are targeting at 5 W for smartphone charging and 15 W for fast charging and laptop charging.

At the early phase of the consortium/alliance, WPC only supported inductive power transfer, and A4WP only supported resonant WPT. However, to cater for a larger market, the two standards are now supporting both operating modes. There are three main differences between these two mainstream standards. First, the Qi devices operate with a frequency range of 87–205 kHz, while the AirFuel devices operate at an ISM band of 6.78 MHz $\pm$ 15 kHz. Second, Qi employs a multi-coil transmitter structure or magnet-core auto-alignment for horizontal freedom and electromagnetic interference (EMI)/electromotive force (EMF) reduction, while AirFuel only uses a larger transmitter coil for three-dimensional (3D) freedom. Third, Qi uses load modulation (also known as back-scattering) for data communication between the WPT receiver and transmitter; while AirFuel uses the Bluetooth module for data communication and power control.

Besides the above relatively mature near-field standards, there is one more emerging wireless charging technology that uses the WiFi frequency bands (2.45 or 5.8 GHz) for longer WPT distance (within a range of 10 m) and total spatial freedom with transmission power of 1 W [24]. This technology is attractive especially for office-room and in-car applications. To achieve effective power transmission, a complex and costly phased array of antenna is required to precisely send power to the device-under-charging with multiple energy beams. This technology is also included in the uncoupled power transfer family of the AirFuel Alliance.

1.3 Design Perspectives

1.3.1 Comparison of Wired and Wireless Power

In the circuit designer's perspective, as shown in Fig. 1.4, wired power is "vertically" provided to the functional circuits from the power supply to ground that is "orthogonal" to the signal lines. In such a case, the power-management circuit designers may treat the signal processing circuits as a black box, and the analog and mixed-signal circuit designers similarly may treat the power supply circuits as a black box. On the other hand, wireless power is "horizontally" transmitted from the energy source to the end-users. In other words, it is "in parallel" with the wireless signal paths. Now, a circuit designer needs to know both power processing and signal processing in order to design an efficient wireless power transfer system. In addition, knowledge of coil design and EM wave analysis is definitely helpful. In fact, in many cases both the wireless power link and the data telemetry link are designed by the same circuit designer. This imposes design challenges to both the circuit and the system designers.

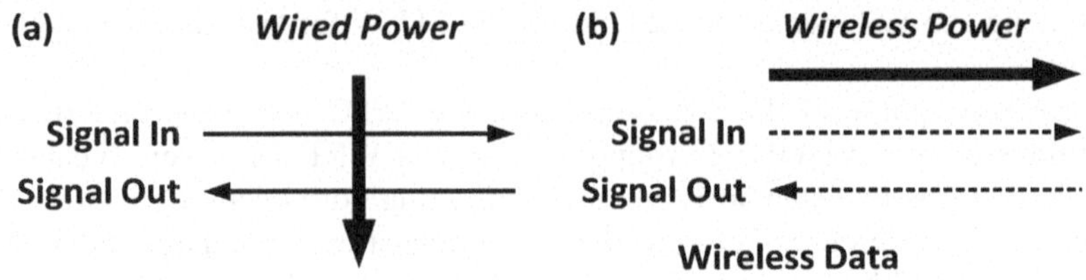

Fig. 1.4 A conceptual diagram showing systems with wired power and wireless power

1.3.2 System Overview

Figure 1.5 shows a typical integrated wireless power transfer system with a power link, a data up-link and a data down-link. The transmitter (TX) chip consists of a DC-DC converter that supplies power to the power amplifier (PA) driving the primary resonator, a load-current detection block that sends information to the microprocessor for output power control, and a data TX for sending information to the receiver (RX) chip. The RX chip consists of a rectifier that converts the AC power into a DC power reservoir, a DC-DC converter that regulates the output voltage or current to the load, a microcontroller that commands data communication, and protection circuits that prevent over-temperature, over-voltage and over-current conditions. Output power control can be realized by supply modulation of the PA [25].

As mentioned in Sect. 1.2.1, increasing the transmission frequency could significantly reduce the values and thus the volume of passive components. Obviously, higher frequency operation enables full integration of a WPT system to within a few cubic millimeters [26–28]. One additional advantage of increasing the WPT frequency is that the data coil/antenna may be removed by transmitting up-link and down-link data using the same power link in a time-multiplexing fashion.

To sum up, for different WPT applications, the engineers need to consider the tradeoffs among various parameters such as transmission frequency, process technology, resonator topology, output regulation method etc. that makes the design very challenging. In addition, considerations for different power levels are quite different. Techniques and topologies that can increase the link efficiency and reduce the device volume are highly desirable. We believe that there are plenty of research opportunities in this area, and we believe consumers would enjoy a smart living environment with minimum wires dangling around.

1.4 Organization of the Book

This book focuses on the circuit and system designs of wireless power transfer for portable and medical devices. The organization of the materials is as follows.

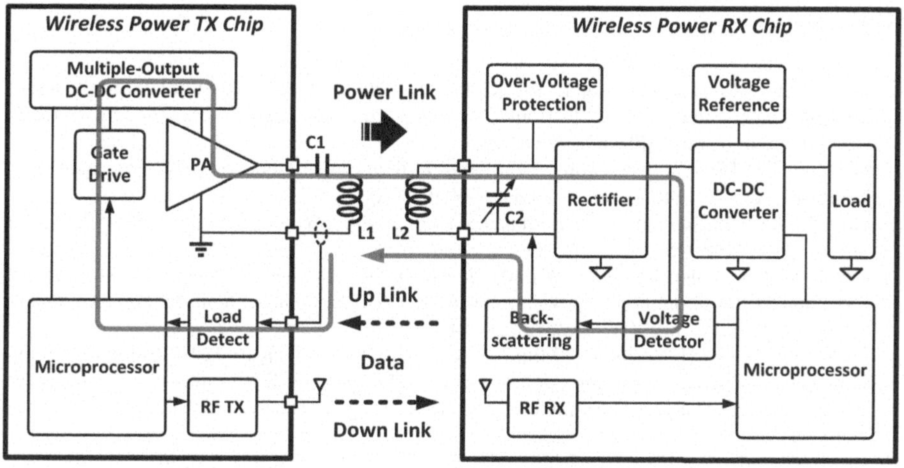

Fig. 1.5 Block diagram of a generic inductive power link with up and down data links

Chapter 1 introduces WPT applications, standards, and some high-level considerations.

Chapter 2 reviews the state-of-the-art WPT systems with the emphasis on reducing the device volume. In recent years, many system architectures have been proposed by various research groups with improved efficiency, reduced number of components and enhanced functionalities.

Chapter 3 analyzes the coil coupling characteristics, and provides design guidelines for optimizing the wireless power link.

Chapter 4 gives a detailed review on CMOS rectifiers for both near-field and far-field operations that enable high-efficiency full-chip integration. Active rectifiers with different delay compensation schemes are introduced and compared with design examples. CMOS rectifier designs for RF energy harvesting are discussed as well.

Chapter 5 discusses the design guidelines for linear regulators, which are the most widely adopted voltage regulation method on-chip, due to its simple structure and area-efficient and noise-reduced characteristics. Fast local control loops can be easily achieved for linear regulators, without affecting the global loop of wireless power transfer system. Two fully-integrated design examples are presented to verify the design suggestions.

Chapter 6 presents the design considerations on switching DC-DC converters which can be classified as switched-inductor and switched-capacitor converters. Converter architectures and control-loop designs for both types are introduced.

Chapter 7 compares the two most popular power amplifiers (Class-D and Class-E) for wireless power transmitters. Essential knowledge of power devices and power electronics for switching loss reduction is also revisited.

Chapter 8 concludes the topic with suggestions on future directions.

References

1. Wise KD, Anderson DJ, Hetke JF, Kipke DR, Najafi K (2004) Wireless implantable microsystems: high-density electronic interfaces to the nervous system. Proc IEEE 92:76–97. doi:10.1109/JPROC.2003.820544
2. Weiland JD, Liu W, Humayun MS (2005) Retinal prosthesis. Annu Rev Biomed Eng 7:361–401. doi:10.1146/annurev.bioeng.7.060804.100435
3. Gosselin B (2011) Recent advances in neural recording microsystems. Sensors 11:4572–4597. doi:10.3390/s110504572
4. Lee H-M, Ghovanloo M (2012) An adaptive reconfigurable active voltage doubler/rectifier for extended-range inductive power transmission. IEEE Trans Circuit Sys II Exp Briefs 59:481–485. doi:10.1109/TCSII.2012.2204840
5. Radecki A, Chung H, Yoshida Y, Miura N, Shidei T, Ishikuro H, Kuroda T (2011) 6W/25mm^2 inductive power transfer for non-contact wafer-level testing. In: IEEE international solid-state circuits conference digest of technical papers (ISSCC), pp 230–232. doi:10.1109/ISSCC.2011.5746297
6. Chung H, Radecki A, Miura N, Ishikuro H, Kuroda T (2012) A 0.025–0.45 W 60%-efficiency inductive-coupling power transceiver with 5-bit dual-frequency feedforward control for non-contact memory cards. IEEE J Solid State Circ 47:2496–2504. doi:10.1109/JSSC.2012.2206686
7. Lu Y, Huang M, Cheng L, Ki WH, SP U, Martins RP (2017) A dual-output wireless power transfer system with active rectifier and 3-level operation. IEEE Trans Power Electron 32:927–930. doi:10.1109/TPEL.2016.2601623
8. (2006) IEEE standard for safety levels with respect to human exposure to radio frequency electromagnetic fields, 3 kHz to 300 GHz. IEEE Std C951–2005 (Revision of IEEE Std C951–1991) 1–238. doi:10.1109/IEEESTD.2006.99501
9. Gabriel S, Lau RW, Gabriel C (1996) The dielectric properties of biological tissues: II. Measurements in the frequency range 10 Hz to 20 GHz. Phys Med Biol 41:2251. doi:10.1088/0031-9155/41/11/002
10. Finkenzeller K (2010) RFID handbook: fundamentals and applications in contactless smart cards, radio frequency identification and near-field communication, 3rd edn. Wiley, Hoboken
11. Merlin R (2007) Radiationless electromagnetic interference: evanescent-field lenses and perfect focusing. Science 317:927–929. doi:10.1126/science.1143884
12. Kim S, Ho JS, Poon ASY (2013) Midfield wireless powering of subwavelength autonomous devices. Phys Rev Lett 110:203905. doi:10.1103/PhysRevLett.110.203905
13. Lu Y, Ki W-H (2014) A 13.56 MHz CMOS active rectifier with switched-offset and compensated biasing for biomedical wireless power transfer systems. IEEE Transac Biomed Circuits Syst 8:334–344. doi:10.1109/TBCAS.2013.2270177
14. Lee H-M, Ghovanloo M (2012) An adaptive reconfigurable active voltage doubler/rectifier for extended-range inductive power transmission. IEEE Transac Circuits Sys II Exp Briefs 59:481–485. doi:10.1109/TCSII.2012.2204840
15. Lu Y, Li X, Ki W-H, Tsui C-Y, Yue CP (2013) A 13.56MHz fully integrated 1X/2X active rectifier with compensated bias current for inductively powered devices. In: IEEE international solid-state circuits conference digest of technical papers (ISSCC), pp 66–67. doi:10.1109/ISSCC.2013.6487639
16. Jow UM, Ghovanloo M (2010) Optimization of data coils in a multiband wireless link for neuroprosthetic implantable devices. IEEE Transac Biomed Circuits Syst 4:301–310. doi:10.1109/TBCAS.2010.2049491
17. Lidow A, de Rooij M (2014, May) Performance evaluation of enhancement-mode GaN transistors in Class-D and Class-E wireless power transfer systems. Bodo Mag:56–60
18. Riehl PS, Satyamoorthy A, Akram H et al (2015) Wireless power systems for mobile devices supporting inductive and resonant operating modes. IEEE Transac Microwave Theory Tech 63:780–790. doi:10.1109/TMTT.2015.2398413

19. Hwang JT, Lee DS, Lee JH et al (2016) An all-in-one (Qi, PMA and A4WP) 2.5W fully integrated wireless battery charger IC for wearable applications. In: IEEE international solid-state circuits conference digest of technical papers (ISSCC), pp 378–380
20. Meng M, Kiani M (2016) Design and optimization of ultrasonic wireless power transmission links for millimeter-sized biomedical implants. IEEE Transac Biomed Circuits Syst:1–10. doi:10.1109/TBCAS.2016.2583783
21. Ulrich WD (1974) Ultrasound dosage for nontherapeutic use on human beings – extrapolations from a literature survey. IEEE Trans Biomed Eng BME-21:48–51. doi:10.1109/TBME.1974.324362
22. van Schuylenbergh K, Puers R (2009) Inductive powering. Springer, Dordrecht
23. Magnetic resonance and magnetic induction – what is the best choice for my application? https://www.wirelesspowerconsortium.com/data/downloadables/1/2/4/6/magnetic-resonance-or-magnetic-induction.pdf. Accessed 19 Oct 2016
24. Branscombe M (2013) Wireless power: could Cota make it long-distance and mainstream? http://www.qiwireless.com/wireless-power-cota-make-long-distance-mainstream. Accessed 19 Oct 2016
25. Li X, Lu Y, Tsui C-Y, Ki W-H (2014) An adaptive wireless powering and data telemetry system for optic nerve stimulation. In: 2014 I.E. international symposium on circuits and systems (ISCAS), pp 1404–1407. doi:10.1109/ISCAS.2014.6865407
26. O'Driscoll S, Poon ASY, Meng TH (2009) A mm-sized implantable power receiver with adaptive link compensation. In: IEEE international solid-state circuits conference digest of technical papers (ISSCC), pp 294–295. doi:10.1109/ISSCC.2009.4977424
27. Raju S, Li X, Lu Y et al (2014) Efficient wireless power transmission technology based on above-CMOS integrated (ACI) high quality inductors. In: IEEE international electron devices meeting (IEDM), pp 12.4.1–12.4.4. doi:10.1109/IEDM.2014.7047038
28. Zargham M, Gulak PG (2015) Fully integrated in-chip coil in 0.13 CMOS for wireless power transfer through biological media. IEEE Transac Biomed Circuits Syst 9:259–271. doi:10.1109/TBCAS.2014.2328318

Chapter 2
Wireless Power Transfer Systems

Abstract This chapter discusses wireless power transfer (WPT) at the system level, with detailed analyses on state-of-the-art WPT output voltage regulation topologies. Possible combinations of the WPT building block configurations are investigated, compared and summarized. Several novel architectures for efficient WPT were proposed in recent years to reduce the number of passive components as well as to improve the system efficiency or flexibility, and these schemes are reviewed and discussed in this chapter.

Keywords Wireless power transfer • Voltage regulation • Inductive power transfer • DC-DC converter • Power amplifier • Rectifier • Coupled coils

2.1 Introduction

A WPT system contains nearly all kinds of power converters, including a DC-AC converter (inverter, serving as a non-linear power amplifier), a transformer (tightly or loosely coupled WPT coils), an AC-DC converter (rectifier), a DC-DC converter and/or a linear regulator. A generic WPT system including most of the building block combinations is shown in Fig. 2.1.

The WPT system consists of a power transmitter (TX) and a power receiver (RX) for transferring power through an inductive power link; a data down-link for sending commands from the primary side to the secondary side; and a data up-link for transferring data from the secondary side to the primary side, for output voltage regulation, for example. The power amplifier (PA) drives the primary series or parallel L_1C_1 resonator, and the secondary series or parallel L_2C_2 resonator receives the wireless power, and then feeds the AC power to the rectifier. The traditional rectifier is a passive diode-bridge that drives a large filtering capacitor, producing an unregulated DC voltage with substantial output voltage ripples. The rectifier is usually cascaded with a voltage regulator such as a linear series/shunt regulator and/or a DC-DC converter. The linear regulator can only provide an output voltage that is lower than the peak voltage of the rectifier, because its output voltage is controlled by tuning the on-resistance of the power transistor. On the other hand, a DC-DC converter consists of energy-storing components (inductors and/or capacitors) and switches may step-up or step-down its output voltage. One should be

Y. Lu, W.-H. Ki, *CMOS Integrated Circuit Design for Wireless Power Transfer*,
Analog Circuits and Signal Processing, DOI 10.1007/978-981-10-2615-7_2

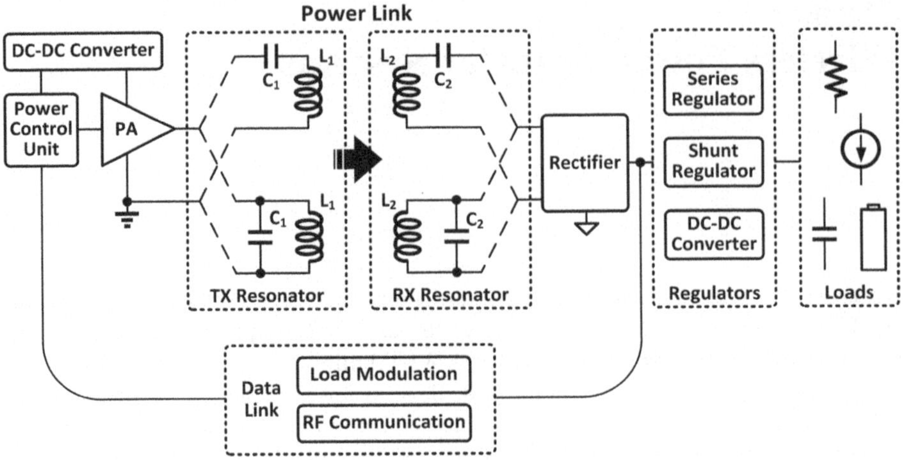

Fig. 2.1 Block diagram of a generic inductive wireless power transfer system

aware that the passive components of the DC-DC converter occupy considerable chip and/or printed circuit board (PCB) area. Moreover, the switching operation of a DC-DC converter generates large output voltage ripples and ground noises. Design considerations and tradeoffs of each building block will be introduced in details in subsequent chapters.

At the receiver side, voltage regulation can be done locally using a linear regulator or a DC-DC converter as discussed above, or globally by adjusting the power from the transmitter side. One global voltage regulation scheme is to feed the digitized rectifier output voltage back to the primary side through load modulation (backscattering, for example) or out-band RF communication (Bluetooth, for example). When the digitized rectifier output voltage is read and decoded by the power control unit (PCU) on the primary side, the PCU will then change the PA output power accordingly. If a Class-D PA is used, the duty ratio will be adjusted; or if a Class-E PA is used, the power supply voltage will be adjusted through a DC-DC switching converter.

The total power conversion efficiency η_{TOTAL} of a WPT link is the product of the efficiencies of each stage, as given by

$$\eta_{TOTAL} = \eta_{DC\text{-}DC} \times \eta_{PA} \times \eta_{LINK} \times \eta_{RECT} \times \eta_{REG} \tag{2.1}$$

where $\eta_{DC\text{-}DC}$ and η_{PA} are the efficiencies of the DC-DC converter and the PA on the TX side, η_{LINK} is the efficiency of the coupling link, and η_{RECT} and η_{REG} are the efficiencies of the rectifier and the regulators on the RX side, respectively. The global voltage regulation loop consists of a few power processing blocks, and is slow compared to the RX local voltage regulation loop, but global power control is essential as it has a major impact on the efficiency of the whole system. It is because when the received power is more than that demanded by the load the excessive power will be consumed by the linear regulator or other over-voltage protection circuits. In such a case, η_{REG} will be very low. Let each stage achieves a reasonably good efficiency of 90%, then the total efficiency is only 59%, which cannot be considered as high. Nevertheless, it would still be beneficial if the efficiency of one certain stage is increased by 2%, and η_{TOTAL} can then be increased by 1.18%. Thus, efficiency optimization for each stage is important.

As depicted in Fig. 2.1, the inductive coupling coils on both the primary inductor and the secondary inductor can be connected either in series or in parallel with the resonance capacitor. Thus, there are four types of resonance combinations for the inductive power link: series-to-series, series-to-parallel, parallel-to-series, and parallel-to-parallel. The selection of resonance type for high system efficiency depends on the load current level and the coupling coefficient that would affect the reflected equivalent secondary impedance on the primary side of the WPT system [1–3]. Detailed analyses of the coupled coils will be given in Chap. 3.

The load of a WPT system could be mixed-signal/digital circuits that are commonly modeled as a resistor in parallel with a filtering capacitor, or analog circuits or neuron stimulation electrodes that can be modeled as a current source, or a rechargeable battery that is similar to a super capacitor. The regulator design and the system design should fully take the load characteristics into consideration, and a specific regulation circuit should be designed for a specific load [4, 5]. Sometimes a high output voltage (say >10 V) is needed for a specific application, for example, the neuron stimulation of a biomedical implant or the flash memory write operation of a wirelessly powered data storage device. A large winding ratio between L_1 and L_2 can be used to boost up the RX output voltage, eliminating one additional step-up DC-DC converter in the RX side.

2.2 Output Voltage Regulation Schemes

The load on the secondary side is driven by a well-regulated voltage source or current source. One simple scheme is as follows: the PA of the TX side always transmits its maximum power, while the output of the rectifier is regulated by a shunt regulator (such as a zener diode) to bypass the excessive power to ground. Alternatively, a DC-DC converter or a linear regulator could be cascaded to the rectifier stage to provide an accurately regulated output voltage [6–9]. In many cases, multiple supply voltages are needed for different load circuits, for example, 0.5–1 V for low-power digital signal processing circuits, 1.2–1.8 V for analog circuits, and 10–20 V for neuron stimulation or memory writing. Therefore, multiple-output topologies have to be studied for the WPT systems. With the PA always transmitting at the maximum power, the system efficiency is low when the power demanded by the load is low. If a global voltage regulation scheme is installed, the up-link data through load-modulation could be back-scattered to the primary side to adjust the PA output power.

An example of a typical inductive power link for implantable medical devices is shown in Fig. 2.2. For the RX chip, a shunt regulator provides coarse voltage regulation and over-voltage protection, and a series-type low-dropout (LDO) regulator provides essential fine voltage regulation for the load circuits. The data

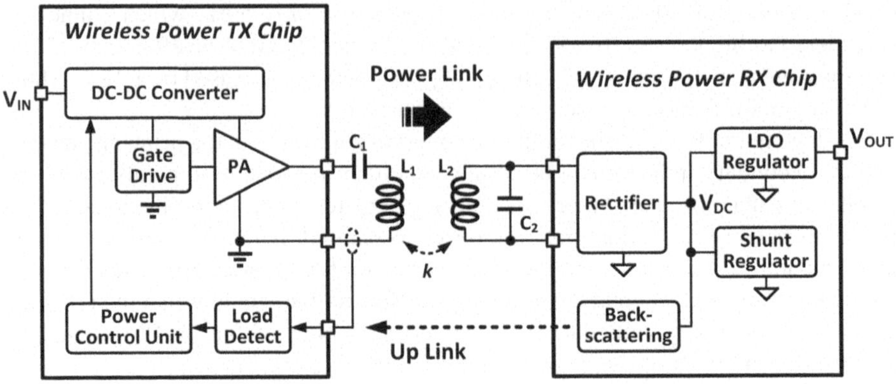

Fig. 2.2 A typical inductive wireless power transfer system with linear regulators and data up-link through load modulation for power control

up-link is realized by back-scattering through changing the input impedance of the rectifier (by shorting out the two AC inputs, for example) to transmit digitized data of the voltage information V_{DC} back to the primary side using the power-link frequency carrier. In this case, the RX side needs more and larger energy-storing components, as normally no energy can be extracted during the back-scattering durations. The up-link data is recovered by magnetic sensing of the PA output current through a detection coil and a load detector. The data is then used to control the supply voltage of the PA and thus the transmitted output power.

The global voltage-regulation loop consists of several low-frequency poles due to the DC-DC converter, the PA driven inductive link, the rectifier, *etc*. An integrator with a pole at DC, that is, $1/s$, is needed in the power control unit (PCU) to stabilize the global loop. A frequency compensator with one or more zeros may be used in the PCU to improve the phase margin. If a higher WPT frequency and/or a higher DC-DC switching frequency is chosen, smaller passive components can be used in the power converters, and the frequencies of the non-dominant poles can also be higher. Consequently, the stability problem of the global loop can be alleviated, the loop bandwidth can be widened, and the transient response can be faster.

In many state-of-the-art designs, novel system-level output-voltage regulation schemes have been proposed to eliminate the post-conversion stage, and consequently system complexity can be reduced and the total efficiency can be increased [10–26]. These schemes are discussed in the following sub-sections.

2.2.1 *Primary Side Non-linear Power Control*

Voltage regulation can be implemented on the primary side using linear control methods (for example, amplitude modulation, AM) or using non-linear control methods. With TX side power control, regulation blocks in the RX side may be eliminated in principle, and the total system efficiency can be higher. However, the global control loop is usually not fast enough to respond to the load transient on the RX side, and the local regulation blocks should not be removed. In one linear control method, the supply voltage of the PA may be changed to tune the output power [6, 7], because the on-resistance of the switches is almost linearly proportional to its gate-to-source voltage V_{GS}. To change the PA supply voltage linearly and consequently the output power, the speed is limited by the large passive components of the DC-DC converter and the resonance power link. Thus, for a faster transient response to the load, primary side non-linear control methods may be used [10–13] as discussed below.

Figure 2.3 shows a primary side non-linear control method by changing the PA switching frequency between the LC resonance frequency F_{RES} and its odd-order sub-harmonic frequency $F_{RES}/3$ [11]. In this system, the energy delivered from the PA to the LC tank is reduced from once every resonance period at the input frequency of F_{RES} to once every 3 resonance periods at the input frequency of $F_{RES}/3$, while the amplitude of the PA output voltage is kept constant, as shown by the conceptual waveforms in Fig. 2.3. Thus, the transmitted powers at these two different input frequencies are changed. Note that although the input frequency of the PA would vary between F_{RES} and $F_{RES}/3$, the LC resonance tanks of the power link still always operate at their resonant frequency, because the resonance tanks are band-pass filters.

The hysteretic comparator on the RX side serves as a one-bit analog-to-digital converter (ADC), and the digitized RX power demand signal is obtained by comparing the RX output voltage with a reference voltage, and fed back to the TX side by the data up-link. At the TX side, the digitized power demand signal is

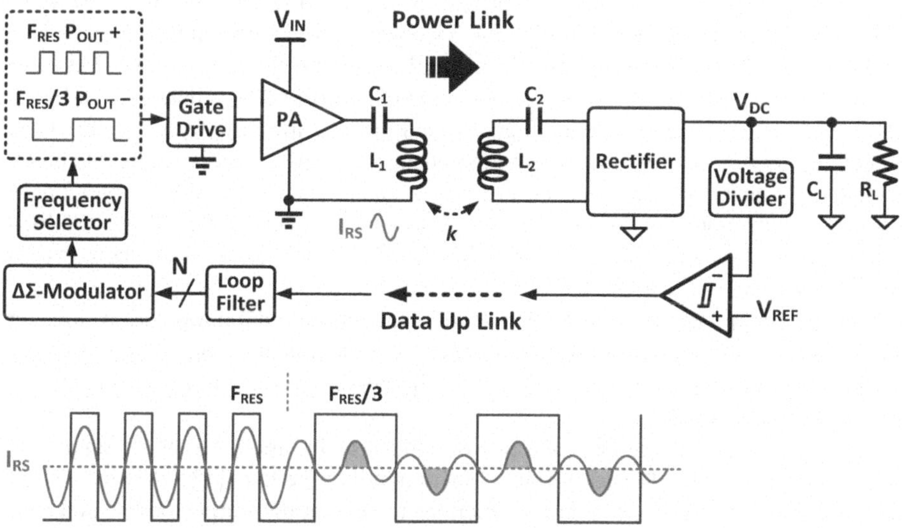

Fig. 2.3 A WPT system with output power controlled by transmission frequency hopping

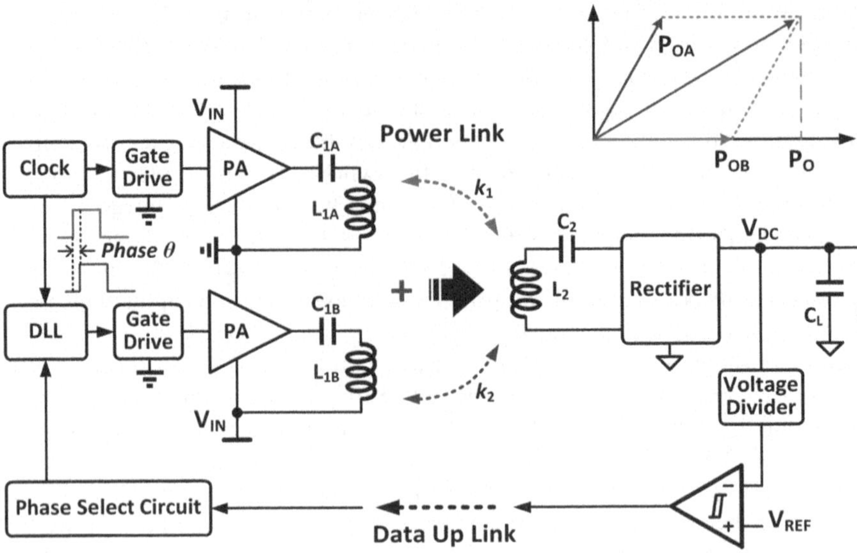

Fig. 2.4 A WPT system with output power controlled by the phase difference between two transmitters

integrated by a digital loop filter and processed by a ΣΔ-modulator for frequency domain noise-shaping. A frequency-control signal is then generated for spurious emission reduction. The power level is adjusted through mode-switching between F_{RES} and $F_{RES}/3$, and higher power means the F_{RES}-mode would occur more frequently than the $F_{RES}/3$-mode.

A two-transmitter topology was proposed in [12] also for non-linear primary-side power control, as shown in Fig. 2.4. Two physically overlapping TX coils L_{1A} and L_{1B} are driven by two separate PAs, and the input clocks of these two PAs have the same frequency but different phases, and are controlled by a delay-locked loop (DLL). When the power demand signal is fed back to the primary side, the phase select circuit will increase or decrease the phase difference θ between the two clock signals to adjust the combined output power of the two PAs. With larger phase difference, the combined output power is lower, and vice versa. The combined output power P_O can be approximately obtained as $P_{OA} \times (1 + \cos\theta)$ where P_{OA} is the output power of a single PA.

One problem with this two-PA structure is that, although the regulation stage on the RX side is removed for higher total efficiency, the transmission efficiency is low as the transmitted power is reduced while the power loss of the transmitter is not reduced proportionally.

Another advanced primary-side power control technique is to set the duty ratio of the operation of the PA according to the demand of the load [13]. When a "Stop" signal comes in, the PA will stop switching to reduce the output power and thus reduce the switching loss at the same time. In [13], the PA stops periodically for a fixed duration of time (which is also called constant-idle or constant-off time control), while the operating time varies with the load condition.

2.2.2 *Reconfigurable Rectifier for Adaptive Output*

In a WPT system with no primary-side power control, a reconfigurable rectifier that can adapt its operating mode according to the output voltage level can provide coarse regulation instead and extend the power transmission range as demonstrated in [14–16]. Figure 2.5 shows an inductively coupled WPT system with a reconfigurable rectifier for low-power applications. A hysteretic comparator is used to compare the rectifier DC output voltage with the reference voltage, and to decide the operating mode of the reconfigurable rectifier. The hysteresis window used in Fig. 2.5 is relatively large (1 V), and amounts to only 1-bit resolution, and V_{DC} is not accurate enough for driving the load. Thus, an LDO regulator is still needed in such a system for accurate regulation.

A typical 1X/2X reconfigurable rectifier with passive diodes is also shown in Fig. 2.5. When the mode-switch M_1 is turned off, the four diodes operate as a full-wave rectifier, denoted as the 1X mode. When M_1 is turned on, the two diodes D_1 and D_2 along with the two capacitors C_{L1} and C_{L2} then operate as two half-wave

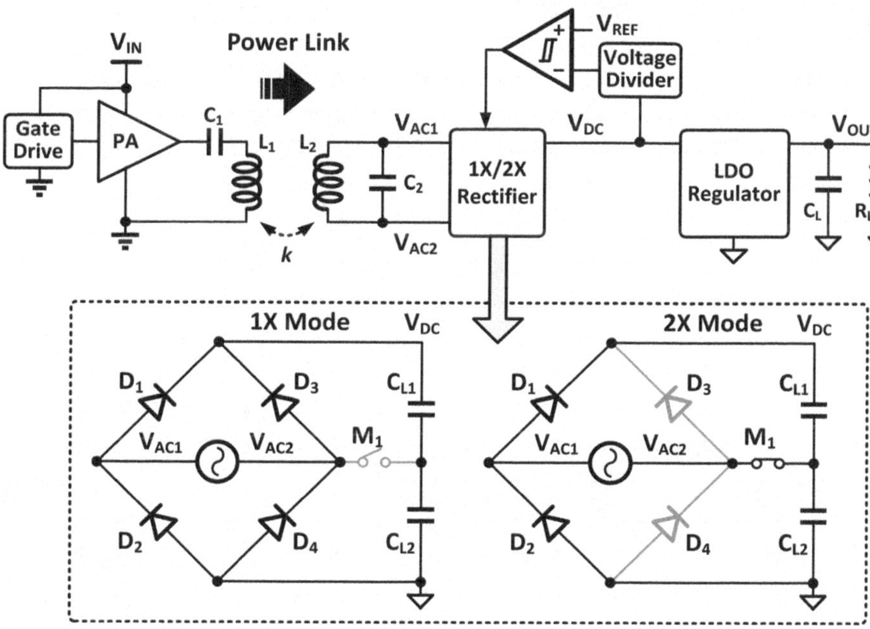

Fig. 2.5 A WPT system with reconfigurable rectifier for coarse regulation and transmission distance extension

rectifiers stacked together to form a voltage doubler, denoted as the 2X mode. The other two diodes stay off in this mode, and the voltage V_{AC2} is clamped to a DC potential roughly equals to half of V_{DC}.

Assume that the same AC input amplitude and the same load resistance are maintained, and assume that the power conversion is ideal with unity efficiency. The 2X mode will provide an output voltage that is two times that of the 1X mode, and with the same load resistance, the load current is also doubled. Hence, the power sourced from the AC input is four times larger, and the equivalent input impedance of the 2X mode seen by the AC input is 1/4 that of the 1X mode. This impedance characteristic should be taken into account when designing the system operating point. However, the analysis is not as straightforward, as the assumption of the AC amplitude remained unchanged under mode change basically does not hold in a WPT system, because the load, be it load current or load resistance, will change in different modes, and so as the equivalent impedance reflected to the primary side. Consider the general case illustrated in Fig. 2.6a. The load resistor on the RX side is mapped to the primary side as $R_{L,EQ}$, and the PA has an equivalent output impedance of R_S. As shown in Fig. 2.6b, when $R_{L,EQ}$ is equal to R_S, the output power delivered to the equivalent load P_L is the maximum, and the ideal efficiency at this point is 50%. We learn that $R_{L,EQ}$ in 2X mode is smaller than that in 1X mode, and if $R_{L,EQ}$ in 1X mode happens to be smaller than R_S (on the left side of the curves), the available output power will decrease when the rectifier changes to 2X mode with an even smaller $R_{L,EQ}$. Thus, to make the reconfigurable rectifier scheme works properly, the system has to operate on the right side of the curves (at low power levels). This limits the scheme to be valid only for low-power applications, such as implantable medical devices.

Besides, the input reactance of the reconfigurable rectifier will also vary under different operating modes. Thus, an adaptive input matching network that can tune the resonance capacitor automatically would be helpful for achieving higher efficiency and higher output power.

By considering all the above factors, a reconfigurable rectifier is proposed to provide an alternative power control strategy for extending the output power level and transmission distance, and increasing the system efficiency.

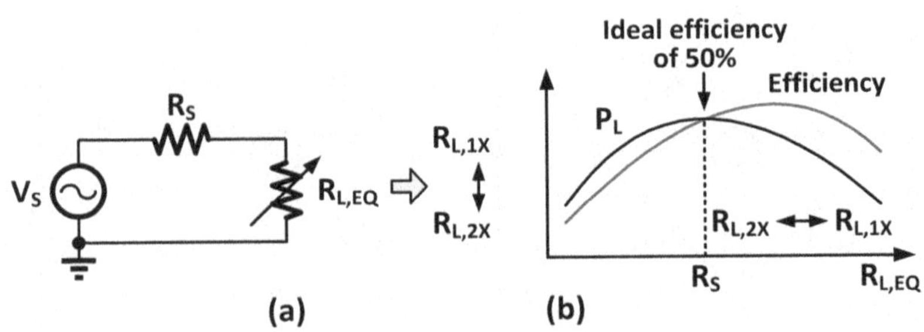

Fig. 2.6 (**a**) Simple equivalent circuit model of a WPT system, and (**b**) illustration of the relationship of load impedance, output power, and efficiency

2.2.3 Resonant Regulating Rectifier

Wireless power transfer systems can be classified into two categories: inductive power transfer (IPT) and resonant wireless power transfer (R-WPT). Both schemes generate a magnetic field by a resonant circuit to transmit power. An IPT system requires the coupling coils be placed closely together and well-aligned for high efficiency, and a magnetic core is often used. It can operate over a wide range of transmission frequency, and as a low-Q system, it eases the requirements on the passive components. For an R-WPT system, the coils can be loosely coupled such that there is more spatial freedom for the related application. Both the primary coil and the secondary coil are tuned with either a series or a parallel capacitor to form LC resonant tanks with very high Q. Both coils are tuned to the same resonant frequency such that the power transmission distance can be extended with a reasonable efficiency. However, the resonant operation is sensitive to component variations, since the LC resonant tank has high Q, and thus the transmitted power spectrum is narrow band, and the efficiency is relatively lower than an IPT system because of the weaker coupling.

Actually, an R-WPT system is pretty much like the topology of a conventional resonant DC-DC converter system, with the transformer in the middle being replaced by a pair of loosely coupled coils. When the switching frequency of the resonant converter exactly coincides with the resonant frequency, the resonant converter operates optimally with reduced switching loss, and consequently provides high power conversion efficiency. But when the resonant converter is optimized at one operating point, it is hard to control the output power. Besides, in R-WPT applications such as mobile phone charging or powering implantable medical devices, it is difficult to achieve the resonant condition with cm-sized or mm-sized coils at a low frequency in the kHz range, and a resonant frequency in the MHz range is required instead.

A time-domain control method was proposed to control a resonant DC-DC converter in [17] by adjusting the output voltage while still satisfying the optimum operating conditions. By selectively enabling or disabling the converter at integral multiples of half cycles, the output power can be tuned. A similar idea was applied to control the power of an R-WPT system as well, and is discussed below.

Let us consider that the receiver side employs series resonance as shown in Fig. 2.7. The resonator composed of L_2 and C_2 acts as an AC current source that supplies current to the rectifier. A switch M_1 with pulse-width modulation (PWM) control is added to the rectifier output to control the output voltage, and is regarded as a one-switch resonant regulating rectifier (3R) in [17]. When M_1 is on, the resonator is designed to operate in its resonating mode, so the resonant current I_{RS} increases and is delivered to the load, and V_{OUT} goes up. When M_1 is off, no current goes to the load from the rectifier and the output capacitor is drained by the load and V_{OUT} decreases, the resonator deviates from its resonating point and I_{RS} decreases. Thus, by tuning the turn-on duty cycle of M_1, the output voltage can be regulated without using an additional power converter or linear regulator. The

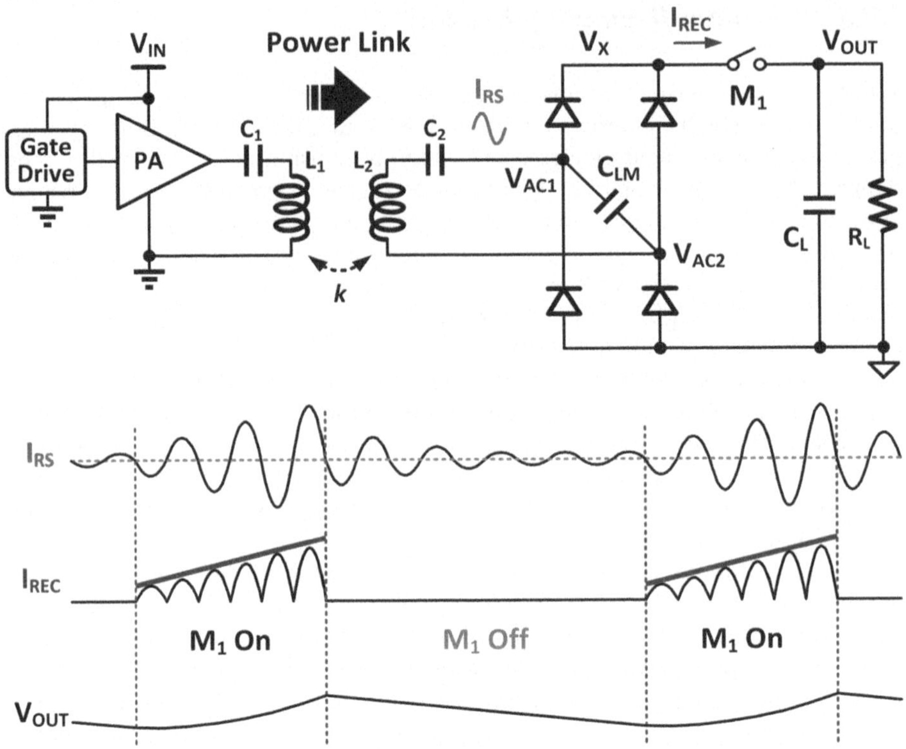

Fig. 2.7 A resonant WPT system with one-switch pulse-width modulation regulating rectifier

limiting capacitor C_{LM} is used to reduce the voltage peak on the node V_X, thus preventing the over-voltage condition when M_1 is in the off state. The resonant regulating rectifier is designed to operate in the discontinuous conduction mode (DCM). It delivers current to the load only discontinuously and results in large peak current of I_{REC} that degrades the efficiency, as the conduction loss of M_1 is given by

$$P_{\text{Cond.Loss}} = \frac{1}{N \cdot T} \int_{t_0}^{t_0+N \cdot T} I_{\text{REC}}^2(t) \cdot R_{\text{ON}} \text{d}t, \tag{2.2}$$

where R_{ON} is the total on-resistance of M_1 and the diodes, T is the operating period and N is an integer. Thus, reducing the peak current of I_{REC} is favorable for increasing the power conversion efficiency.

Consequently, the continuous conduction mode (CCM) operation was also proposed in [17] with one flying capacitor C_{FLY} and three switches M_1–M_3, as shown in Fig. 2.8. The flying capacitor is switched to stack on top of the output capacitor like a step-down charge pump to provide charge for the load. When M_1 and M_3 are off and M_2 is on, C_{FLY} is in series with C_L. Hence, V_{OUT} is roughly ½

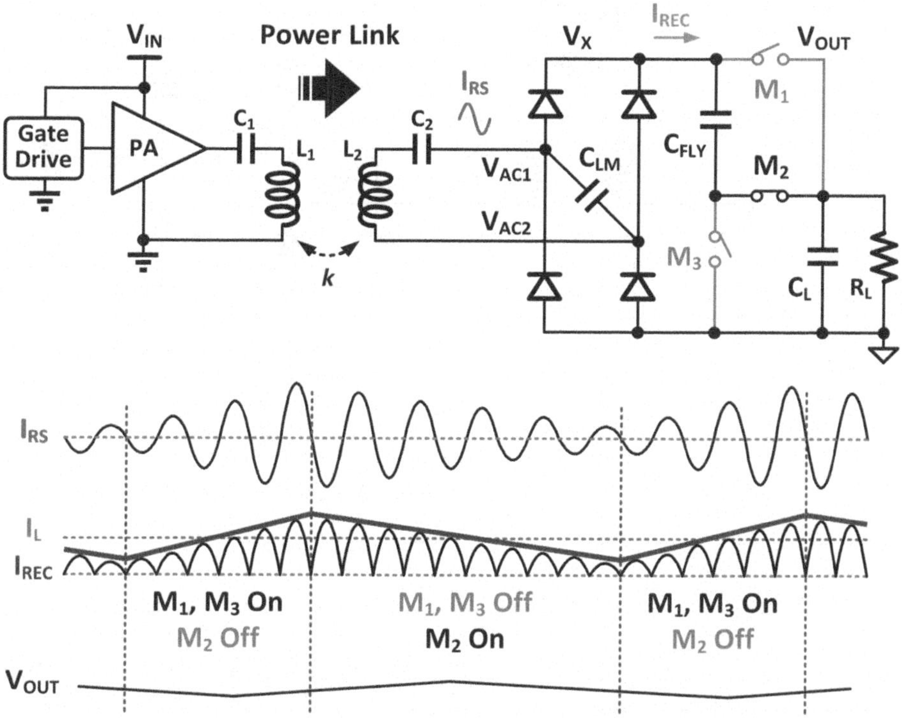

Fig. 2.8 A resonant WPT system with pulse-width modulation regulating rectifier and step-down charge pump

times of V_X, and I_{REC} starts to decrease. When M_1 and M_3 are on and M_2 is off, C_{FLY} is in parallel with C_L, and the full-wave rectifier charges up V_{OUT}. In CCM operation, I_{REC} only drops to zero temporarily and immediately rebounds, and hence provides a continuous current to the load, and results in higher power conversion efficiency and lower output voltage ripples. Moreover, the current variations of I_{RS} and I_{REC} are not as pronounced, relieving the voltage stress on V_X, and consequently smaller C_{LM} could be used.

2.2.4 Reconfigurable Resonant Regulating Rectifiers

Based on the knowledge of the above two sub-sections, a 1X/2X reconfigurable resonant regulating (R^3) rectifier was designed in [19] as illustrated in Fig. 2.9. The reconfiguration switch M_1 is now controlled by a PWM signal as well. When M_1 is off, the rectifier operates in 1X mode and the ideal output in this case is $V_{AC,Peak}$. When M_1 is on, the rectifier operates in 2X mode and tends to reach to 2 times of $V_{AC,Peak}$. Thus, when the rectifier is switched between 1X mode and 2X mode,

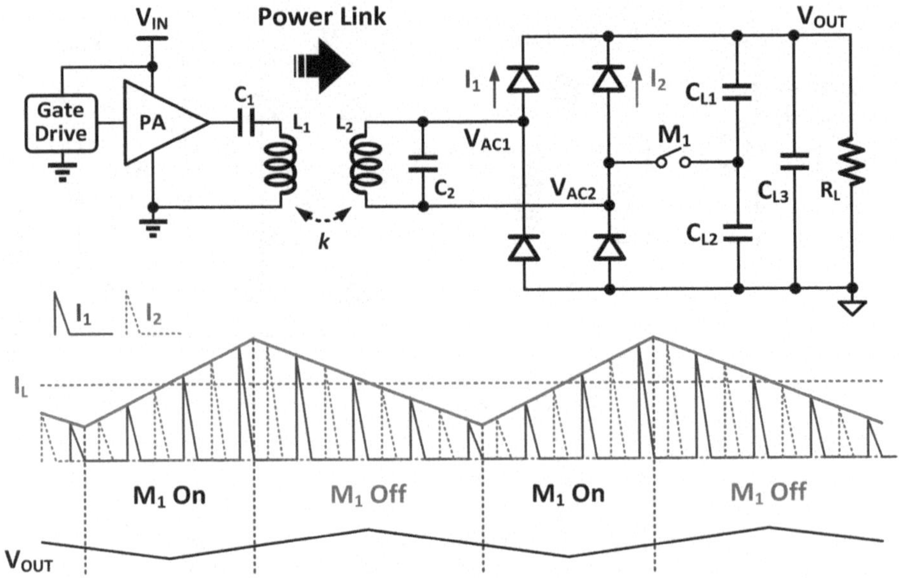

Fig. 2.9 A WPT receiver with 1X/2X regulating rectifier with pulse-width modulation

V_{OUT} decreases and increases periodically and can be regulated at the prescribed voltage level. The waveforms plotted in Fig. 2.9 show the CCM case; and of course, the regulating rectifier can operate in DCM at light-load conditions as well. Hence, the output can be well regulated under different load conditions.

Another advanced version of the R^3 rectifier was proposed in [20] as shown in Fig. 2.10. This reconfigurable rectifier has three modes: 1X, ½X, and 0X modes. In the 1X mode, both M_1 and M_2 are off, and the active diodes operate as a full-wave rectifier that can provide large output current to the load. In the ½X mode, M_1 is on and M_2 is off, V_{AC2} is shorted to ground. Then, only one diode out of the four will conduct current I_1 to the load as a half-wave rectifier. Compared to the 1X mode, the current delivered to the load is halved, and this mode is named as the ½X mode. In the 0X mode, both M_1 and M_2 are on, and the receiver resonant tank L_2 and C_2 is freewheeling. This is different from the M_1 off-state of the one-switch 3R operation, whence the rectifier is disconnected from the load, and breaks the path of the inductor current.

Output regulation of the above topology can be achieved by switching between two of the three modes, to achieve a finer regulation. In heavy load conditions, mode-switching between 1X and ½X modes are activated; and in light load conditions, ½X and 0X modes are activated. Therefore, CCM operations can be maintained for a wide load current range. A smooth transition between different modes during load transient can be achieved with accurate load current sensing techniques which add minor circuit complexity to the on-chip implementation.

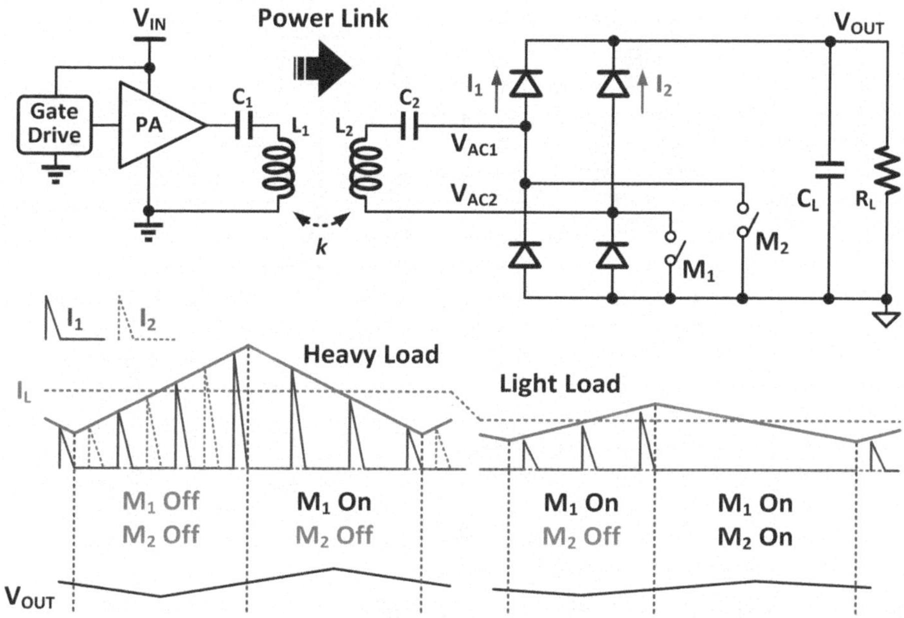

Fig. 2.10 A WPT receiver with 1X/½X/0X regulating rectifier with pulse-width modulation

2.2.5 *Pre-rectifier Regulation*

Besides the above mentioned primary-side power control and reconfigurable and/or regulating rectifier topologies for output regulation, pre-rectifier regulation topology was proposed in [21, 22], as shown in Fig. 2.11. Similar to the previous mode-switching schemes using PWM signal, the pre-rectifier regulation scheme (named as Q-modulation in [21]) modulates the load impedance as seen from the resonator, and achieves output regulation.

As shown by the conceptual waveforms in Fig. 2.11, the switch M_1 is turned on at the zero-crossing point of the resonating current I_{RS}, which is commonly known as zero-current switching (ZCS). To avoid the switch from experiencing high voltage and high current simultaneously that causes unnecessary conduction loss, ZCS is an important operation principle to achieve high efficiency. During the on-state of M_1, the high-Q resonant tank L_2 and C_2 stores the maximum energy that is transferred from the primary side. During the off-state of M_1, the energy stored in the resonant tank together with the continually transferred energy is delivered to the load through I_1 and I_2 of the rectifier. Therefore, the amount of energy that is transferred to the load can be controlled by tuning the duty cycle D of the switch. Different from the mentioned regulation schemes, this scheme achieves regulation within one cycle at the resonant frequency and requires faster control circuits (faster transistors).

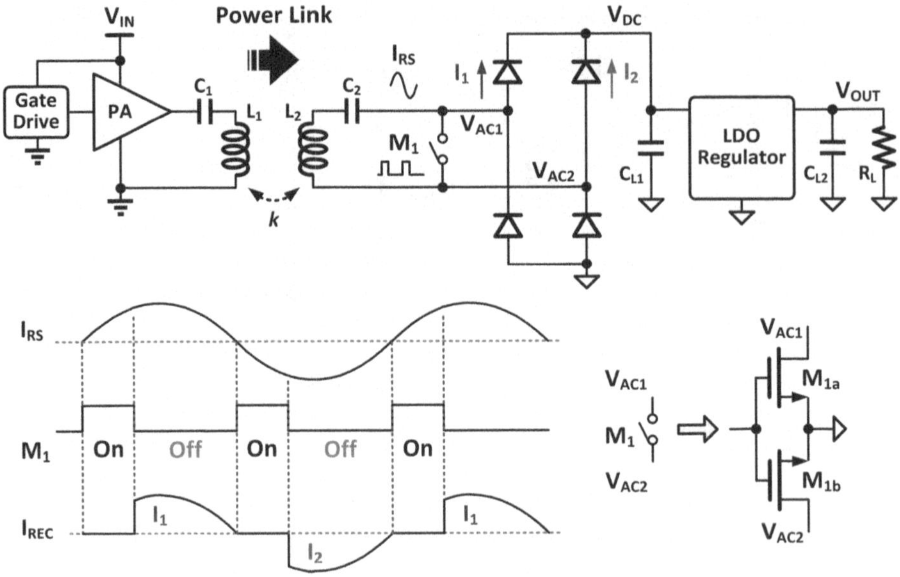

Fig. 2.11 A WPT receiver with PWM pre-rectifier regulation

The maximum input amplitude (the carrier envelope) is tracked by a dedicated feedback loop as designed in [21], such that the duty cycle D will be increased or decreased accordingly. Since the loop only tracks the input voltage, an LDO regulator is still needed to obtain an accurate output voltage. On the other hand, in [22], the rectifier output was used to compare with a reference voltage to generate the PWM signal. Therefore, the post-stage LDO regulator can be removed.

In fact, this topology and its operation principle are similar to a resonant DC-DC boost converter, in which the inductor accumulates energy first and then delivers the energy to the load through a diode. The switch M_1 can be implemented by an NMOS transistor with the gate-drive referenced to the ground potential, as shown in Fig. 2.11. The equivalent load impedance $R_{L,EQ}$ seen from the resonator can be roughly computed as

$$R_{L,EQ} = \frac{(8/\pi^2)R_L}{M^2} = (1-D)^2(8/\pi^2)R_L, \tag{2.3}$$

where $M = 1/(1-D)$ is the voltage conversion ratio of the boost converter operating in CCM. The full-wave rectifier converts the impedance R_L into $(8/\pi^2)\,R_L$, assuming that the LDO regulator is ideal. With a PWM feedback loop, the Q-modulation technique can transform the load impedance automatically for any load and coupling variations.

2.2.6 Multi-Level Single-Inductor Multiple Output Operation

In many applications, different power supply levels are required by the system for various functions [26, 27], while the ripples generated by the converters should be reasonably small for noise-sensitive loads. Therefore, the bulky passive components of the converters should be utilized effectively [15] or even be reused with time multiplexing. For an application that requires multiple outputs, single-inductor multi-output (SIMO) techniques could be employed to reduce the number of inductors [27, 28]. A multi-level SIMO operation for the WPT RX was proposed in [26], which merged a multi-level SIMO switching converter with a multi-stage rectifier as shown in Fig. 2.12.

In such configuration, any output voltage that has a value between V_{DCk} and $V_{DC(k-1)}$ ($k = 1, 2, \ldots, N$), can be regulated by sinking currents from those two rectifier outputs through L_3 as a buck converter. The voltage across L_3 can be smaller than that of the typical two-level operation, such that the inductor current and the output voltage ripples are reduced because of the reduced voltage swing across L_3. Moreover, for the voltage boosting application, the AC input is boosted without introducing the well-known right-half-plane zero that exists in boost and buck-boost converters, which means that the loop bandwidth of the buck converter can be designed at higher frequency for fast transient responses. It is also noted that the maximum achievable efficiency of the series-parallel multi-stage rectifier does not change with the number of stages, since all the stages get energy from the AC input in one phase and then being stacked to attain a high V_{DC} in the complementary phase.

To demonstrate the idea, a 3-level single-inductor dual-output DC-DC converter that inherently cooperates with a 2X active rectifier (voltage doubler) that gives two DC outputs (3-level) was designed for the WPT RX [26], as shown in Fig. 2.13.

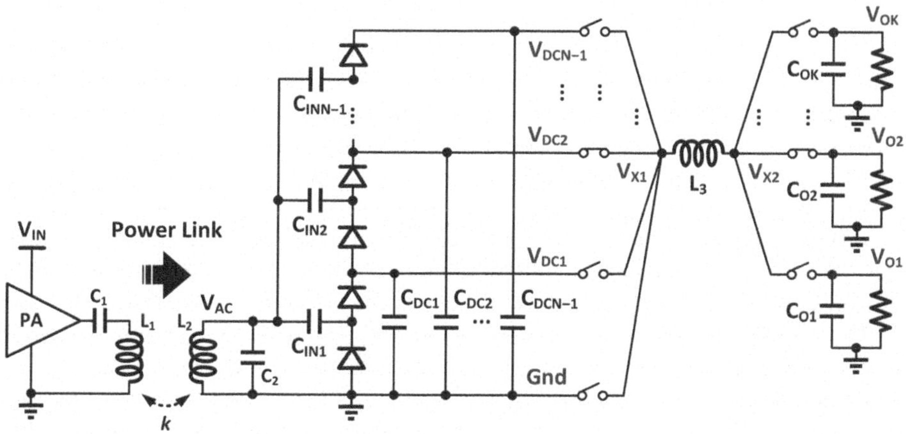

Fig. 2.12 A WPT receiver with a 2(N − 1)-stage rectifier merged with N-level single-inductor multi-input multiple-output converter

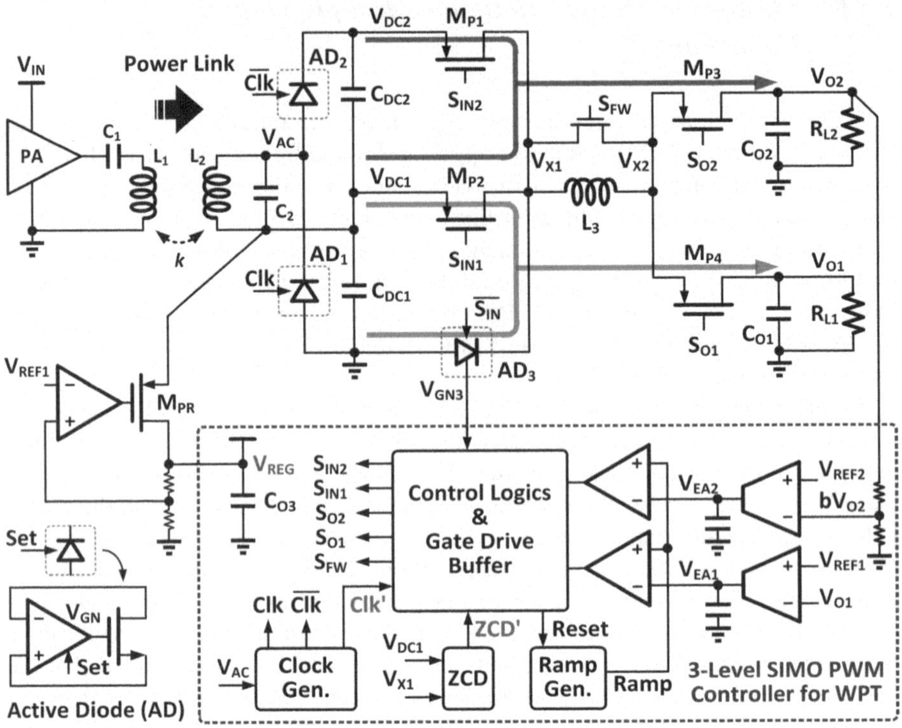

Fig. 2.13 A WPT system with 2X rectifier and 3-level SIMO converter

Assume the V_{AC} amplitude is 3.2 V, V_{DC1} will be around 2.7 V and V_{DC2} will be approximately 5.5 V. The higher output voltage V_{O2} retrieves current from V_{DC2} and V_{DC1}, and can be programmed to range from 3.3 V to 5 V for I/O and memory circuits; while the lower output voltage V_{O1} retrieves current from V_{DC1} and Gnd, and can be programmed to range from 1.0 V to 1.8 V for core circuits.

The SIMO converter also switches at the WPT frequency of 6.78 MHz. One benefit of operating the DC-DC converter at the WPT frequency is that, part of the discontinuous rectifier output currents will directly go to the DC-DC converter inputs in every half cycle, bypassing the rectifier load capacitors C_{DC1} and C_{DC2}, thus further reduces the output ripple. Now, the volume of a passive component is roughly proportional to its value. In this work, smaller L_3 and capacitors can be used, benefiting from the 3-level and 6.78 MHz discontinuous conduction mode (DCM) operation. An independent PWM control loop is used for each output, such that cross-regulation can be reduced as long as the converter works in DCM.

The timing diagram of the 3-level SIMO converter in DCM operation is shown in Fig. 2.14. The clock pulse Clk' that initiates the PWM control for V_{O1} is recovered from V_{AC} by an inverter and a pulse generator, and the zero current detection (ZCD) signal that determines the connection of V_{X2} is generated by sensing the current of M_{P2}. Since both V_{O2} and V_{O1} need to get current from

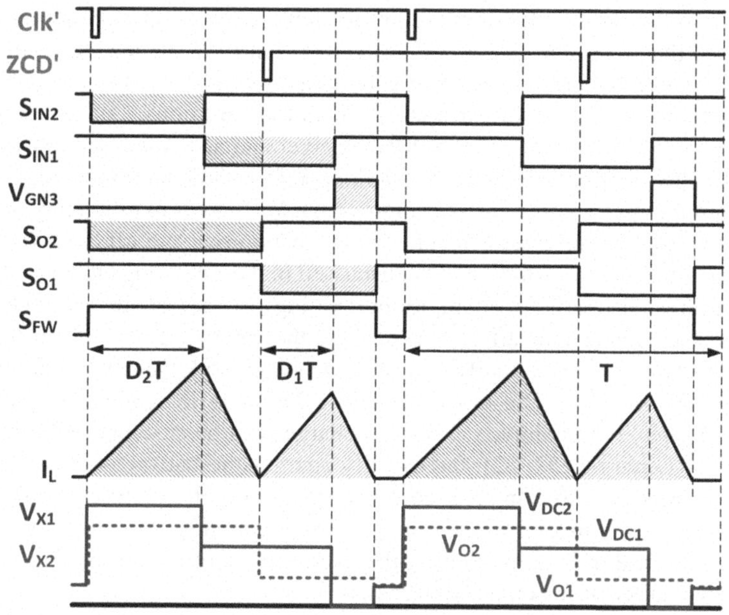

Fig. 2.14 Timing diagram of the 3-level SIMO converter in DCM operation

V_{DC1}, M_{P2} is kept on during the output transition period to avoid one-time switching loss of M_{P2} in every cycle. A comparator is used to detect the V_{X1} voltage crossing point. When $V_{X1} > V_{DC1}$ during the on-state of S_{IN1}, a pulse ZCD' that initiates the PWM control for V_{O2} will swap the output control signals S_{O2} and S_{O1}. A large quiescent current is needed to increase the speed of the comparator so as to reduce the reverse current of M_{P2}. Alternatively, the speed requirement on the ZCD can be relaxed by using an additional slow auto-calibration loop to adjust the comparator offset, as demonstrated in [3]. Since the calibration loop only requires a low bandwidth, for example 200 kHz in [3], its current consumption is only on the order of 1 μA. A freewheel switch S_{FW} will be turned on at the end of each cycle when both AD_3 and M_{P4} are off, to suppress the possible ringing caused by L_3 and the parasitic capacitors at the V_{X1} and V_{X2} nodes when they are floating [28]. For higher conversion efficiency and/or large power handling capability of the SIMO DC-DC converter, lower switching frequency can be used [29].

2.3 Summary and Discussion

For wireless power transfer systems, as reviewed in this chapter at the system level, output voltage regulation can be achieved by using primary side power control, reconfigurable/regulating rectifiers, as well as pre-rectifier regulation topologies, or simply cascading a DC-DC converter stage in the receiver. For the regulating rectifiers and the pre-rectifier regulation schemes, pulse-width modulation is commonly used that is similar to DC-DC converter control of an accurate output. The equivalent load impedance is automatically tuned by the dedicated PWM loop that adapts to the load and coupling distance/orientation variations. Consequently, high efficiencies can be achieved for the WPT receiver and also the entire WPT system.

Besides the mentioned equivalent load impedance modulation techniques, there are two more direct ways to further improve the WPT system efficiency. One is to achieve the goal at the circuit level, that is, to improve the efficiencies of each cascading power stages, which will be discussed in the following chapters. Take the active rectifier design as an example, optimizing the size of the cross-connected transistors can enhance the efficiency, because the parasitic capacitors of the cross-connected transistors are actually part of the resonant tank that should not be taken as switching loss [30]. The other feasible solution is to reduce the number of converter stages. For example, a wireless charger can be designed without a post-stage DC-DC charger [31, 32], but directly charges the battery or super-capacitor through the rectifier only. In such case, the system regulation loop should control the charging current instead of the output voltage. One stage of conversion loss can thus be eliminated. However, sensing the averaged charging current for regulation would be a challenge for circuit designers.

In terms of device volume, bulky passive components of the AC-DC and DC-DC converters should be utilized effectively or be reused with time multiplexing for miniaturized devices. In addition, operating at a higher WPT frequency can significantly reduce the sizes of the passive components. Moreover, smaller inductance with higher Q can be more easily achieved at higher frequencies within a limited space, which is one of the key factors for improving the WPT link efficiency.

Last but not least, similar to all other analog circuit designs, a dedicated feedback loop controlling the critical parameter is essential for a good WPT design. Meanwhile, for robust operation, a WPT system that has multiple loops, the dynamics of each local (fast) and global (slow) loop has to be analyzed carefully and thoroughly.

References

1. Wang C-S, Covic GA, Stielau OH (2004) Power transfer capability and bifurcation phenomena of loosely coupled inductive power transfer systems. IEEE Trans Ind Electron 51:148–157. doi:10.1109/TIE.2003.822038
2. Ahn D, Hong S (2014) Wireless power transmission with self-regulated output voltage for biomedical implant. IEEE Trans Ind Electron 61:2225–2235. doi:10.1109/TIE.2013.2273472
3. Cheng L, Ki WH, Lu Y, Yim TS (2016) Adaptive on/off delay-compensated active rectifiers for wireless power transfer systems. IEEE J Solid State Circuits 51:712–723. doi:10.1109/JSSC.2016.2517119
4. Lee H-M, Park H, Ghovanloo M (2013) A power-efficient wireless system with adaptive supply control for deep brain stimulation. IEEE J Solid State Circuits 48:2203–2216. doi:10.1109/JSSC.2013.2266862

5. Lee H-M, Ghovanloo M (2013) A power-efficient wireless capacitor charging system through an inductive link. IEEE Transac Circuits Syst II Express Briefs 60:707–711. doi:10.1109/TCSII.2013.2278104
6. Wang G, Liu W, Sivaprakasam M, Kendir G (2005) Design and analysis of an adaptive transcutaneous power telemetry for biomedical implants. IEEE Transac Circuits Syst I Regul Pap 52:2109–2117. doi:10.1109/TCSI.2005.852923
7. Li X, Lu Y, Tsui C-Y, Ki W-H (2014) An adaptive wireless powering and data telemetry system for optic nerve stimulation. In: IEEE international symposium on circuits and systems (ISCAS), 2014, pp 1404–1407. doi:10.1109/ISCAS.2014.6865407
8. Park H, Jang J, Kim H, Park Y, Oh S, Pu Y, Hwang KC, Yang Y, Lee K (2016) A design of a wireless power receiving unit with a high-efficiency 6.78-MHz active rectifier using shared DLLs for magnetic-resonant A4 WP applications. IEEE Trans Power Electron 31:4484–4498. doi:10.1109/TPEL.2015.2468596
9. Moti K-G, Neri F, Moon S, Yeon P, Yu J, Cheon Y, Roh Y, Ko M, Park B-H (2015) 12.9 A fully integrated 6W wireless power receiver operating at 6.78MHz with magnetic resonance coupling. In: IEEE international solid- state circuits conference – (ISSCC), 2015, pp 1–3
10. Radecki A, Chung H, Yoshida Y, Miura N, Shidei T, Ishikuro H, Kuroda T (2011) 6W/25mm^2 inductive power transfer for non-contact wafer-level testing. In: IEEE international solid-state circuits conference digest of technical papers (ISSCC), 2011, pp 230–232
11. Shinoda R, Tomita K, Hasegawa Y, Ishikuro H (2012) Voltage-boosting wireless power delivery system with fast load tracker by ΔΣ-modulated sub-harmonic resonant switching. In: IEEE international solid-state circuits conference digest of technical papers (ISSCC), 2012, pp 288–290
12. Tomita K, Shinoda R, Kuroda T, Ishikuro H (2012) 1-W 3.3–16.3-V boosting wireless power transfer circuits with vector summing power controller. IEEE J Solid State Circuits 47:2576–2585. doi:10.1109/JSSC.2012.2211698
13. Huang C, Kawajiri T, Ishikuro H (2016) A wireless power transfer system with enhanced response and efficiency by fully-integrated fast-tracking wireless constant-idle-time control for implants. In: 2016 I.E. symposium on VLSI circuits (VLSI-Circuits). doi:10.1109/VLSIC.2016.7573491
14. Lee H-M, Ghovanloo M (2012) An adaptive reconfigurable active voltage doubler/rectifier for extended-range inductive power transmission. IEEE Transac Circuits Syst II Express Briefs 59:481–485. doi:10.1109/TCSII.2012.2204840
15. Lu Y, Li X, Ki W-H, Tsui C-Y, Yue CP (2013) A 13.56MHz fully integrated 1X/2X active rectifier with compensated bias current for inductively powered devices. In: 2013 IEEE international solid-state circuits conference (ISSCC), pp 66–67. doi:10.1109/ISSCC.2013.6487639
16. Lee H (2016) An auto-reconfigurable 2×/4× AC-DC regulator for wirelessly powered biomedical implants with 28% link efficiency enhancement. IEEE Transac Very Large Scale Integ (VLSI) Syst 24:1598–1602. doi:10.1109/TVLSI.2015.2452918
17. Joung GB, Rim CT, Cho GH (1989) Integral cycle mode control of the series resonant converter. IEEE Trans Power Electron 4:83–91. doi:10.1109/63.21875
18. Choi J-H, Yeo S-K, Park S, Lee J-S, Cho G-H (2013) Resonant regulating rectifiers (3R) operating for 6.78 MHz resonant wireless power transfer (RWPT). IEEE J Solid State Circuits 48:2989–3001. doi:10.1109/JSSC.2013.2287592
19. Li X, Tsui C-Y, Ki W-H (2015) A 13.56 MHz wireless power transfer system with reconfigurable resonant regulating rectifier and wireless power control for implantable medical devices. IEEE J Solid State Circuits 50:978–989. doi:10.1109/JSSC.2014.2387832
20. Cheng L, Ki W-H, Wong Y-T, Yim T-S, Tsui C-Y (2016) A 6.78MHz 6W wireless power receiver with a 3-Level 1×/½×/0× reconfigurable resonant regulating rectifier. In: 2016 IEEE international solid-state circuits conference (ISSCC), pp 376–377
21. Kiani M, Lee B, Yeon P, Ghovanloo M (2015) A Q-modulation technique for efficient inductive power transmission. IEEE J Solid State Circuits 50:2839–2848. doi:10.1109/JSSC.2015.2453201

22. Ahn D, Kim S, Moon J, Cho IK (2016) Wireless power transfer with automatic feedback control of load resistance transformation. IEEE Trans Power Electron. doi:10.1109/TPEL.2015.2513060
23. Lee B, Kiani M, Ghovanloo M (2016) A triple-loop inductive power transmission system for biomedical applications. IEEE Transac Biomed Circuits Syst 10:138–148. doi:10.1109/TBCAS.2014.2376965
24. Zhao J, Yao L, Xue RF et al (2016) An integrated wireless power management and data telemetry IC for high-compliance-voltage electrical stimulation applications. IEEE Transac Biomed Circuits Syst 10:113–124. doi:10.1109/TBCAS.2015.2404038
25. Lee KFE (2010) A timing controlled AC-DC converter for biomedical implants. In: 2010 IEEE international solid-state circuits conference – (ISSCC), pp 128–129
26. Lu Y, Huang M, Cheng L et al (2017) A dual-output wireless power transfer system with active rectifier and three-level operation. IEEE Trans Power Electron 32:927–930. doi:10.1109/TPEL.2016.2601623
27. Ma D, Ki W-H, Tsui C, Mok PKT (2003) Single-inductor multiple-output switching converters with time-multiplexing control in discontinuous conduction mode. IEEE J Solid State Circuits 38:89–100. doi:10.1109/JSSC.2002.806279
28. Ma D, Ki W-H, Tsui C (2003) A pseudo-CCM/DCM SIMO switching converter with free-wheel switching. IEEE J Solid State Circuits 38:1007–1014. doi:10.1109/JSSC.2003.811976
29. Jing X, Mok PKT, Lee MC (2011) A wide-load-range constant-charge-auto-hopping control single-inductor-dual-output boost regulator with minimized cross-regulation. IEEE J Solid State Circuits 46:2350–2362. doi:10.1109/JSSC.2011.2162188
30. Lu Y, Ki W-H (2014) A 13.56 MHz CMOS active rectifier with switched-offset and compensated biasing for biomedical wireless power transfer systems. IEEE Transac Biomed Circuits Syst 8:334–344. doi:10.1109/TBCAS.2013.2270177
31. Lazaro O, Rincón-Mora GA (2013) 180-nm CMOS wideband capacitor-free inductively coupled power receiver and charger. IEEE J Solid State Circuits 48:2839–2849. doi:10.1109/JSSC.2013.2280053
32. Huang M, Lu Y, U S-P, Martins RP (2017) A reconfigurable bidirectional wireless power transceiver with maximum current charging mode an 58.6% battery-to-battery efficiency. In: 2017 IEEE international solid-state circuits conference – (ISSCC), pp 376–377. doi:10.1109/ISSCC.2017.7870418

Chapter 3
Analysis of Coupled-Coils

Abstract This chapter introduces a design-oriented analysis of various wireless power links driving a resistive load, which include the ideal coupled-coils, the coupled-coils with parasitic resistors, the coupled-coils with series-resonant primary and series-resonant secondary, and the coupled-coils with series-resonant primary and parallel-resonant secondary. The standard equation-pair of coupled-inductors is employed to analyze the above configurations to obtain (1) the reflected impedance of the secondary circuit detected at the primary side, (2) the link voltage gain and (3) the link efficiency. We label the analysis as design-oriented by presenting the equations with parameters arranged in a way that expose the physical meaning of the results and help with designing the coupled-coils more systematically and effectively.

Keywords Wireless power transfer • Coupled coils • Coupled inductors • Parallel resonant • Series resonant • Link voltage gain • Link efficiency • Reflected impedance

3.1 Introduction

One of the pioneer researches of wireless power transfer for biomedical implants is [1]. This paper did not derive the equations from the first principle, and it is difficult for readers to identify the relations among the several loads R, R_o and R_L. Typographical errors also obscure the reading of the results. For example, Fig. 1 of [1] shows a parallel-resonant secondary circuit, but in Fig. 2 the secondary equivalent circuit becomes a series-resonant circuit. Moreover, the inductor, resistor and capacitor of the secondary circuit are labeled L_1, R_1 and C_1 that should belong to the primary circuit. Similar ambiguities apply to [2], and there is no explanation on why the coupling coefficient k has to be smaller than 0.509, or how $P_i(\omega)$ is computed.

In [3], various configurations of coupled-coils such as the non-resonant coupled-coils, and coupled-coils with juxtaposed series-resonant and parallel-resonant coils for the primary and the secondary have been analyzed and even optimized. It seems that the analysis works have been completed, and we only need to worry about applying the results for design. However, the design equations are shrouded in

Y. Lu, W.-H. Ki, *CMOS Integrated Circuit Design for Wireless Power Transfer*, Analog Circuits and Signal Processing, DOI 10.1007/978-981-10-2615-7_3

symbols that could be difficult to decipher. For example, the term R_2/R_L is easily recognized as the ratio of the parasitic resistor R_2 over the load resistor R_L, but in [3] it is written as Q_L/Q_2 $(=(\omega_o L_2/R_L)/(\omega_o L_2/R_2))$ and the meaning is not readily recognized. If all the quality factors (Q's) are expanded in full the equations again become too complicated to read. One essential ingredient of design-oriented analysis is to group the terms tactically that expose the physical meaning underneath.

It should be noted that the basic analysis of a pair of coupled-inductors has been discussed in undergraduate textbooks such as [4]. The derivation is direct and uncomplicated. It is our aim to work out various configurations of coupled-coils following a straightforward methodology, as will be illustrated in the sections below.

3.2 Coupled-Coils and Modeling

3.2.1 *Ideal Transformer*

Before working on the coupled-coils, let us consider the ideal transformer and its modeling, as shown in Fig. 3.1.

The ideal transformer has turns-ratio of 1:n and with the dot-orientation as shown, the (s-domain) I-V characteristic is described by the following equations:

$$V_2 = nV_1 \tag{3.1}$$

$$I_1 = nI_2 \tag{3.2}$$

The circuit model in the s-domain is shown in Fig. 3.1b, and the input impedance $z_{in}(s)$ can be computed as

$$z_{in}(s) = \frac{V_1}{I_1} = \frac{1}{n^2} Z_L(s) \tag{3.3}$$

Simply put, in the above discussion, the ideal transformer is modeled using one current-controlled current source and one voltage-controlled voltage source. In fact, it is possible to construct other equivalent circuits that substitute each controlled source with a current/voltage-controlled current/voltage source. However, in practice, the one shown in Fig. 3.1b proves to be the most convenient.

3.2.2 *Ideal Coupled-Coils*

Next, Fig. 3.2a shows a pair of ideal coupled-inductors with primary inductance L_1, secondary inductance L_2, and mutual inductance M driving a load resistor R_L. The coil system is described in the time-domain by the equation-pair of the coupled-inductors that drive a resistive load:

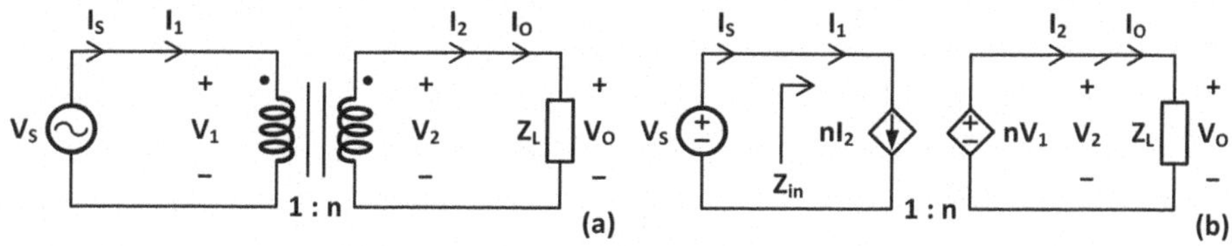

Fig. 3.1 Ideal transformer: (**a**) circuit; and (**b**) circuit model

$$V_1(t) = L_1 \frac{dI_1}{dt} + M \frac{dI_2{}'}{dt} \tag{3.4}$$

$$V_2(t) = M \frac{dI_1}{dt} + L_2 \frac{dI_2{}'}{dt} = -I_2{}' R_L \tag{3.5}$$

The computation would be much simpler when working in the s-domain, and it is beneficial to define $I_2 = -I_2{}'$ as shown in Fig. 3.2b. The corresponding s-domain equations with a generic load impedance Z_L(s) are then given by

$$V_1(s) = sL_1 I_1 - sMI_2 \tag{3.6}$$

$$V_2(s) = sMI_1 - sL_2 I_2 = I_2 Z_L(s) \tag{3.7}$$

To facilitate the computation in different settings, at least three equivalent circuit models have been developed. They are (1) the T-model; (2) the transformer model; and (3) the reflected impedance model. As discussed in Sect. 3.2.1, the transformer model has many variations, and here we only show the most common ones.

3.2.3 T-Model of Coupled-Coils

The T-model is shown in Fig. 3.3, and one limitation is clear: the primary coil and the secondary coil have to share a common ground.

Nevertheless, if we proceed with the derivation, we have

$$V_1(s) = s(L_1 - M)I_1 + sM(I_1 - I_2) \tag{3.8}$$

$$V_2(s) = sM(I_1 - I_2) - s(L_2 - M)I_2 = I_2 Z_L(s) \tag{3.9}$$

The above equations are the same as Eqs. (3.6) and (3.7). The requirement to have a common ground for the primary side and the secondary side defeats the purpose of wireless power transfer, and hence, this model is not useful for our discussion.

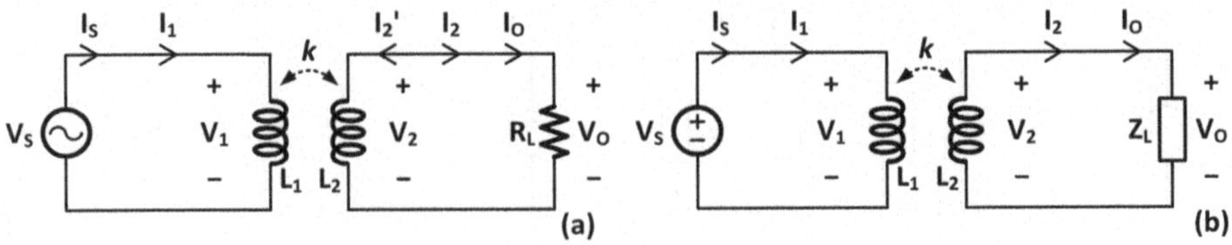

Fig. 3.2 Ideal coupled-coils (**a**) in the time-domain with a resistive load; and (**b**) in the s-domain with a generic load impedance $Z_L(s)$

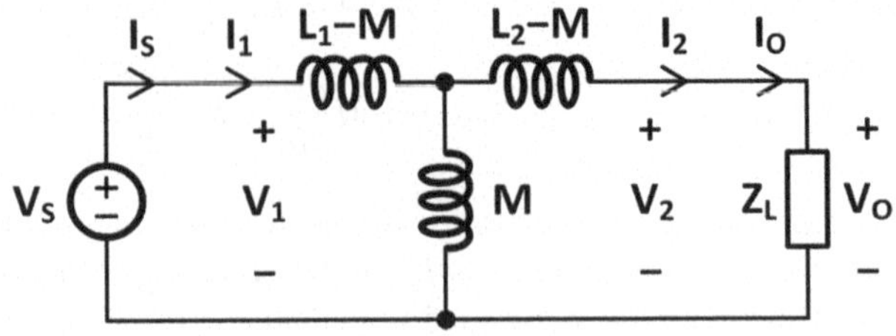

Fig. 3.3 T-model of ideal coupled-coil

3.2.4 *Transformer Model of Coupled-Coils*

The transformer model is shown in Fig. 3.4. The mutual inductance can be accounted for by using the coupling coefficient k and the turns-ratio n that are defined as

$$k = \frac{M}{\sqrt{L_1 L_2}}, n = \sqrt{\frac{L_2}{L_1}} \tag{3.10}$$

The equations of the model can be derived as follows. With reference to Fig. 3.4b, the input voltage $V_1(s)$ and the output voltage $V_2(s)$ are given by

$$V_1(s) = sL_1(I_1 - knI_2) \tag{3.11}$$

$$V_2(s) = knV_1 - s\left(1 - k^2\right)L_2 I_2 \tag{3.12}$$

By substituting k and n of Eq. (3.10) into Eq. (3.11), we have

$$V_1(s) = sL_1 I_1 - sL_1 \frac{M}{\sqrt{L_1 L_2}} \sqrt{\frac{L_2}{L_1}} I_2 \tag{3.13}$$

which is easily seen to be the same as Eq. (3.6). Similarly, with the help of Eq. (3.6), Eq. (3.12) can be rewritten as

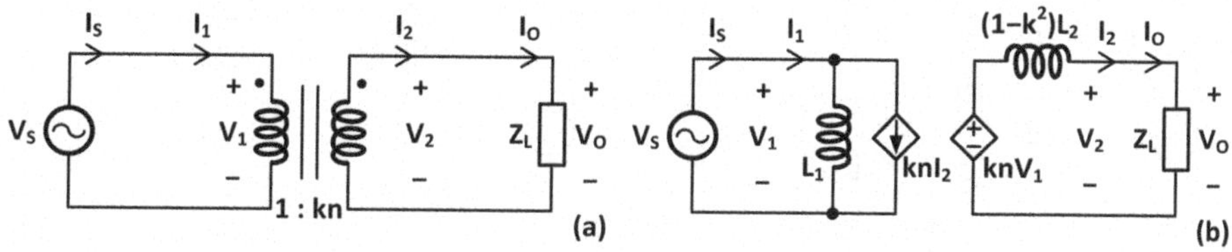

Fig. 3.4 (**a**) Transformer model of ideal coupled-coils, and (**b**) its equivalent circuits on each side

$$V_2(s) = \frac{M}{L_1}(sL_1I_1 - sMI_2) - sL_2I_2 + s\frac{M^2}{L_1L_2}L_2I_2 \tag{3.14}$$

which is the same as Eq. (3.7). Hence, the equation-pairs ((3.5), (3.6)) and ((3.11), (3.12)) are equivalent, and Fig. 3.4b is a valid circuit model of the coupled-coils. As noted in Sect. 3.2.1, there is more than one equivalent model for the transformer, which could be employed to model the coupled-coils, and three models other than the one shown in Fig. 3.4 are discussed in [3].

3.2.5 *Reflected Impedance Model of Coupled-Coils*

In some applications, it may be convenient to incorporate (reflect) the equivalent impedance of the secondary loop to the primary loop for analysis. The reflected impedance model (also known as the equivalent impedance model in [3]) is shown in Fig. 3.5. Note that the modeling of the secondary coil is the same as that of the transformer model of Fig. 3.4, and the advantage of this model is to reflect all components in the secondary side, including the output impedance $Z_L(s)$ and the secondary coil inductance sL_2, to the primary side, such that the two sides can be analyzed independently.

The equations of the primary coil and the secondary coil are easily written down from Fig. 3.5b as

$$V_1(s) = \left(sL_1 + Z_{eq}(s)\right)I_1 \tag{3.15}$$

$$V_2(s) = knV_1 - s\left(1 - k^2\right)L_2I_2 \tag{3.16}$$

To show that Eq. (3.6) is equivalent to Eq. (3.15), we may write I_2 of Eq. (3.6) in terms of I_1 using Eq. (3.7), that is,

$$V_1(s) = sL_1I_1 - sM\left(\frac{sM}{sL_2 + Z_L(s)}\right)I_1 \tag{3.17}$$

With $s = j\omega$ so that $s^2 = -\omega^2$, we define $Z_{eq}(s)$ as

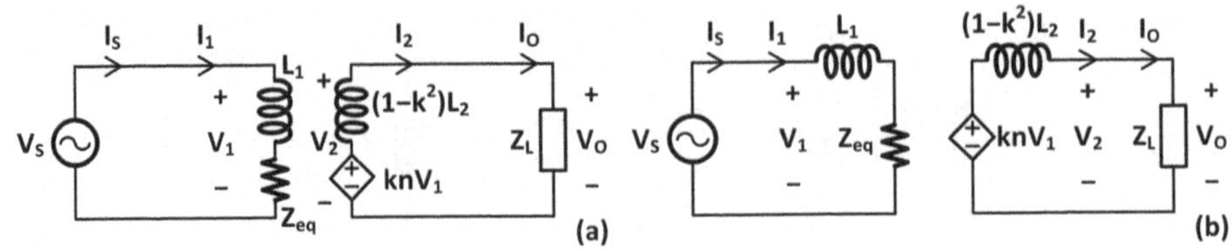

Fig. 3.5 Reflected impedance model. (**a**) Circuit model, and (**b**) model reorganized for clarity

$$Z_{eq}(s) = \frac{\omega^2 M^2}{sL_2 + Z_L(s)} \tag{3.18}$$

and Eq. (3.17) is then the same as Eq. (3.15), and is in turn also equivalent to Eq. (3.6).

3.2.6 Link Voltage Gain and Link Efficiency

One important reason of developing an equivalent circuit model of the coupled-coils is to facilitate the computation of the link voltage gain A_T and the link efficiency η_T in the presence of parasitic components, where the subscript T indicates the complete wireless power link, that is, from the source to the load. These two parameters are defined as

$$A_T = \frac{| V_o |}{V_S} \tag{3.19}$$

$$\eta_T = \frac{P_o}{P_S} \tag{3.20}$$

where V_S and P_S are the source voltage amplitude and the source power at the primary coil, respectively; and $|V_o|$ and P_o are the load voltage amplitude and the load power at the secondary coil, respectively. Note that by construction, V_S is necessarily real.

Computing power in the time-domain involves solving differential equations and evaluating integrals that are daunting tasks. Instead, we assume the circuit to be operating in the sinusoidal steady state, and the computations could then be performed in the phasor domain. Consider a circuit port X with port-voltage $V_x(j\omega)$ and port-current $I_x(j\omega)$, both of which are phasors. The (total) complex power P_{XT} is given by [5]

$$P_{XT} = P_X + jQ_X = \frac{1}{2} V_x(j\omega) I_x(j\omega)^* \tag{3.21}$$

where P_X is the average real power, Q_X and is the average reactive power, and $I_x(j\omega)^*$ is the complex conjugate of $I_x(j\omega)$.

3.2.7 Computing A and η of the Ideal Coupled-Coils

Let us consider the ideal coupled-coils with a resistive load R_L using the reflected impedance (Z_{eq}) model. First, the equivalent impedance of the secondary circuit reflected to the primary side is given by Eq. (3.18), which is equal to

$$Z_{eq}(s) = \frac{\omega^2 M^2}{R_L + sL_2} \tag{3.22}$$

We may define the quality factor of the inductor of the secondary coil driving the load R_L as Q_L:

$$Q_L = \frac{\omega L_2}{R_L} \tag{3.23}$$

and Eq. (3.22) can be written as

$$Z_{eq}(j\omega) = \frac{\omega^2 M^2}{R_L\left(1 + {Q_L}^2\right)}(1 - jQ_L) \tag{3.24}$$

or equivalently,

$$Z_{eq}(j\omega) = \frac{k^2}{n^2}\frac{{Q_L}^2}{\left(1 + {Q_L}^2\right)}R_L(1 - jQ_L) \tag{3.25}$$

The link voltage gain can easily be obtained by referring to Fig. 3.5b as

$$A_T = \frac{|V_o|}{V_S} = \left|knV_S\frac{R_L}{R_L + j\omega(1 - k^2)L_2}\right|\frac{1}{V_s} \tag{3.26}$$

$$\Rightarrow \qquad A_T = \frac{kn}{\sqrt{1 + (1 - k^2)^2 {Q_L}^2}} \tag{3.27}$$

The coupling coefficient cannot be larger than unity, that is, $k \leq 1$; and for the weak coupling case it could be very small, that is, $k < <1$. Moreover, if the load resistance is large then Q_L will not be high, and could even be lower than unity, that is, $Q_L < 1$.

The computation of the link efficiency is more involved. First of all, the average power of the load is given by

$$P_o = \frac{1}{2} V_o I_o{}^* = \frac{1}{2} R_L |I_o|^2 \tag{3.28}$$

Similarly, the complex power of the source P_{ST} is

$$P_{ST} = P_S + jQ_S = \frac{1}{2} V_S I_S{}^* \tag{3.29}$$

In general, the product $V_S I_S{}^*$ cannot be guaranteed to be real, but only the real part will be used to compute the link efficiency. Moreover, the source V_S would be outputting power instead of absorbing power, and to arrive at a positive P_S, I_S is taken as the negative of the port current of V_S instead.

One simple way to compute Eq. (3.29) is to express V_S and I_S in terms of I_2. From Eq. (3.7) (with $Z_L(s) = R_L$) we have

$$I_1 = \frac{R_L + j\omega L_2}{j\omega M} I_2 \tag{3.30}$$

Eq. (3.30) is then substituted into Eq. (3.6) to obtain

$$V_1 = \left(j\omega L_1 \frac{R_L + j\omega L_2}{j\omega M} - j\omega M \right) I_2 \tag{3.31}$$

With $V_S = V_1$, $I_S = I_1$ and $I_o = I_2$, Eq. (3.29) becomes

$$P_{ST} = \frac{|I_o|^2}{2} \left(j\omega L_1 \frac{R_L + j\omega L_2}{j\omega M} - j\omega M \right) \frac{R_L - j\omega L_2}{-j\omega M} \tag{3.32}$$

$$\Rightarrow \quad P_{ST} = \frac{|I_o|^2}{2} \left(R_L - j\omega L_2 + j\omega L_1 \frac{R_L{}^2 + \omega^2 L_2{}^2}{\omega^2 M^2} \right) \tag{3.33}$$

Clearly, the real part of P_{ST} is

$$P_S = \mathrm{Re}(P_{ST}) = \frac{1}{2} R_L |I_o|^2 \tag{3.34}$$

The link efficiency is then equal to unity:

$$\eta_T = \frac{P_o}{P_S} = \frac{|I_o|^2 R_L/2}{|I_o|^2 R_L/2} = 1 \tag{3.35}$$

This result is both encouraging and discouraging. It is encouraging because Eq. (3.35) shows that if the system is ideal and has no parasitic resistance, and even if the coupling is very weak (k << 1), there will be no loss, and this is what is to be expected of an ideal system. It is discouraging exactly for the same reason: it is clear that power is consumed by the primary coil even if there is no load resistance ($R_L = 0\ \Omega$), and Eq. (3.35) simply means that the model is not accurate enough to account for the loss due to electromagnetic radiation. Nevertheless, by recognizing the limitation of a circuit model of the coupled-coils (as compared to using Maxwell equations), we could still proceed to derive the case with parasitic resistance, and obtain some insights of how power is transferred and how power is lost.

3.2.8 Design-Oriented Analysis of Coupled-Coils with Parasitics

Let us consider the coupled-coils driving a resistive load R_L with parasitic coil-resistors R_1 and R_2 as shown in Fig. 3.6. We will follow a similar procedure of computing the link voltage gain A_T and the link efficiency η_T as for the ideal coupled-coils. Design-oriented analysis is adopted in the sense that results are presented in the form that help with designing and optimizing the system.

With reference to Fig. 3.6, the s-domain equations are

$$V_S = R_1 I_1 + V_1 = (R_1 + sL_1)I_1 - sMI_2 \tag{3.36}$$

$$V_2 = sMI_1 - sL_2 I_2 = (R_L + R_2)I_2 \tag{3.37}$$

Let the quality factor of the primary coil L_1 driving its coil-resistor R_1 be Q_1; the quality factor of the secondary coil L_2 driving its coil-resistor R_2 be Q_2; and the quality factor of the secondary coil L_2 driving $(R_L + R_2)$ be Q_S. They are thus given as

$$Q_1 = \frac{\omega L_1}{R_1}, \quad Q_2 = \frac{\omega L_2}{R_2}, \quad Q_S = \frac{\omega L_2}{R_L + R_2} \tag{3.38}$$

Note that for an efficient system the parasitic components should be very small and the quality factors Q_1 and Q_2 should be very high, that is,

$$Q_1, Q_2 >> Q_S \tag{3.39}$$

The reflected (or equivalent) impedance is given by (*c.f.* Eqs. (3.18) and (3.25))

$$Z_{eq}(s) = \frac{\omega^2 M^2}{R_L + R_2 + sL_2} \tag{3.40}$$

$$\Rightarrow \qquad Z_{eq}(j\omega) = \frac{k^2}{n^2}\frac{{Q_S}^2}{1 + {Q_S}^2}(R_L + R_2)(1 - jQ_S) \tag{3.41}$$

To compute the link voltage gain, express I_1 in terms of I_2 using Eq. (3.37) and substitute the result into Eq. (3.36), then

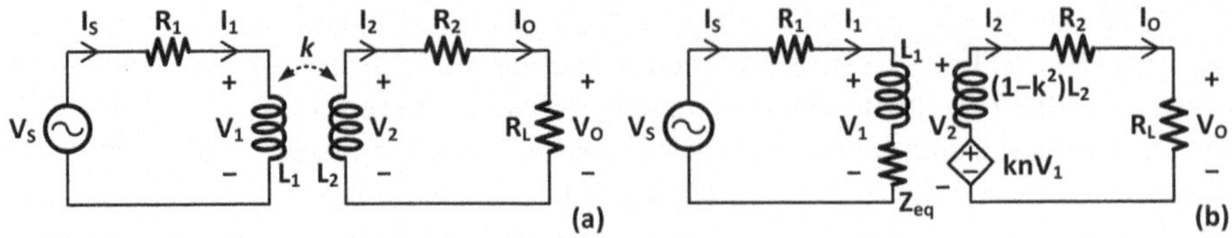

Fig. 3.6 (**a**) The circuit of coupled-coils with parasitic resistances, and (**b**) the reflected impedance model

$$V_S = \left((R_1 + sL_1) \frac{R_L + R_2 + sL_2}{sM} - sM \right) I_o \tag{3.42}$$

Clearly, $|V_o| = |I_o R_L|$, and with $A_T = |V_o(j\omega)|/V_S$, we have

$$A_T = \frac{|\, j\omega M R_L \,|}{|\,(R_1 + j\omega L_1)(R_L + R_2 + j\omega L_2) + \omega^2 M^2\,|} \tag{3.43}$$

The above equation can also be expressed in terms of Q's, and a few lines of arithmetic gives

$$A_T = \frac{knQ_1}{\sqrt{\left(1 - (1 - k^2) Q_1 Q_S\right)^2 + (Q_1 + Q_S)^2}} \frac{R_L}{R_L + R_2} \tag{3.44}$$

We may maximize A_T w.r.t. R_L by working out the following condition:

$$\frac{dA_T}{dR_L} = 0 \tag{3.45}$$

Satisfying Eq. (3.45) requires R_L to be negative, which means that there is no feasible solution to Eq. (3.45). Hence, the minimum and the maximum A_T occurs at the boundary values, that is,

$$A_{T,\min} = A_T|_{R_L=0} = 0 \tag{3.46}$$

$$A_{T,\max} = A_T|_{R_L=\infty} = \frac{knQ_1}{\sqrt{1 + Q_1{}^2}} \tag{3.47}$$

To compute the link efficiency we follow the same procedure as in Sect. 3.2.7 by noting that the load power is $P_o = ½|I_o|^2 R_L$, and I_S can be expressed in terms of I_o in $P_{ST} = ½V_S I_S{}^*$, that is,

$$P_{ST} = \frac{1}{2}\left((R_1 + j\omega L_1)\frac{R_L + R_2 + j\omega L_2}{j\omega M} - j\omega M\right)I_o \times \left(\frac{R_L + R_2 + j\omega L_2}{j\omega M}\right)^* I_o{}^* \tag{3.48}$$

Eq. (3.48) can be manipulated to read

$$P_{ST} = \left(R_L + R_2 + \frac{n^2}{k^2}\left(1 + \frac{1}{Q_S{}^2}\right)R_1 + j\frac{n^2}{k^2}\frac{1 + (1 - k^2)Q_S{}^2}{Q_S{}^2}\omega L_1\right)\frac{|I_o|^2}{2} \tag{3.49}$$

The link efficiency is then given by

$$\eta_T = \frac{P_o}{P_S} = \frac{1}{1 + \frac{n^2}{k^2}\left(1 + \frac{1}{Q_S{}^2}\right)\frac{R_1}{R_L} + \frac{R_2}{R_L}} \tag{3.50}$$

Although this circuit-model derivation cannot account for radiation loss, Eq. (3.50) is very informative. First, let us consider a battery with internal resistance $R_{battery}$ driving a load resistance R_L, and note that $R_{battery}$ is in series with R_L. It is easy to derive that the efficiency is given by

$$\eta = \frac{1}{1 + \frac{R_{battery}}{R_L}} \tag{3.51}$$

Let us consider Eq. (3.50) again. The effect of R_2 on the efficiency is simply similar to that of the internal resistance of a battery. However, the effect of R_1 is much worse. For an implantable medical device (IMD) powered through the coupled-coils, the coupling coefficient k is usually very small, on the order of a few percent, making the term with R_1 very large, thus degrading the link efficiency. Hence, it is of utmost importance to lower the parasitic resistance of the primary coil.

The link efficiency η_T has been worked out in [3] (pp. 48–49). First, in the course of derivation, a parameter R is defined as

$$R = \frac{k^2}{n^2}(R_L + R_2) \tag{3.52}$$

and the link efficiency is derived as

$$\eta_T = \frac{R_L\omega^2 L_1{}^2 k^4}{(R + R_1)\omega^2 L_1{}^2 k^4 + R^2 R_1} \tag{3.53}$$

Next, by substituting Q_L from Eq. (3.23) and Q_1 and Q_2 from Eq. (3.38), η_T is computed to be

$$\eta_T = \frac{1}{1 + \frac{Q_L}{Q_2} + \frac{1}{k^2}\left(\frac{Q_L}{Q_1} + \underline{\frac{1}{Q_1 Q_L}} + \frac{2}{Q_1 Q_2} + \underline{\frac{Q_L}{Q_1 {Q_2}^2}}\right)} \tag{3.54}$$

It can be shown that Eq. (3.54) is equivalent to Eq. (3.50), and the two underlined terms are typographical mistakes in [3] that have been corrected above. Although Eq. (3.54) is correct, it is not as useful as Eq. (3.50), which shows the immediate relationships of R_L, R_1 and R_2.

As a final remark of this section, we allege that the equations Eqs. (3.41), (3.44) and (3.50) are useful for design because important information is revealed. For example, Eq. (3.41) shows how the resistance combination $R_L + R_2$ of the secondary circuit got reflected to the primary side, and how it is being reduced by the coupling coefficient k and the turns-ratio n.

3.3 Resonant Coupled-Coils

For the majority of the wireless power transfer systems, capacitors are added to both the primary-coil and the secondary-coil to resonate at the frequency of power transmission. A capacitor can be added in series with or in parallel to the inductor, and the coil, be it the primary or the secondary, can be made series-resonant or parallel-resonant accordingly. In the majority of cases, the source would be a voltage source V_S that drives a series-resonant circuit. Hence, in the following analysis, the primary side is assumed to have series resonance only.

3.3.1 Series-Series Resonant Coupled-Coils

If a capacitor is added in series with each of the primary-coil and the secondary-coil, then we have a pair of series-series resonant coupled-coils (S-S coils), as shown in Fig. 3.7.

With reference to Fig. 3.7, the s-domain equations are

$$V_S = (1/sC_1 + R_1 + sL_1)I_1 - sMI_2 \tag{3.55}$$

$$V_2 = sMI_1 - sL_2I_2 = (R_2 + 1/sC_2 + R_L)I_2 \tag{3.56}$$

The equivalent impedance is given by

$$Z_{eq}(j\omega) = \frac{\omega^2 M^2}{R_L + R_2 + 1/j\omega C_2 + j\omega L_2} \tag{3.57}$$

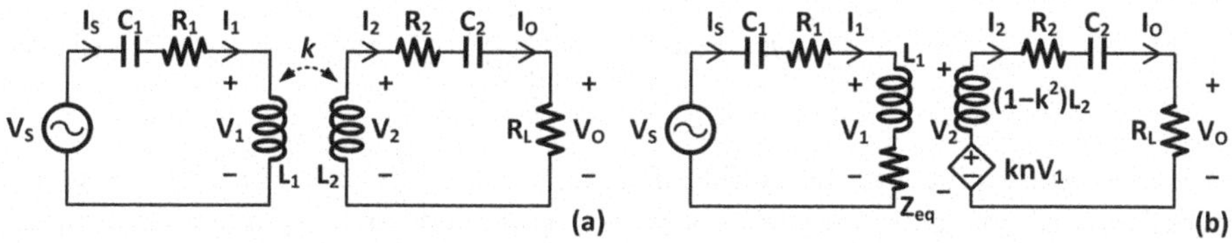

Fig. 3.7 (**a**) The circuit of series-series resonant coupled-coils with parasitic resistances, and (**b**) its reflected impedance model

It is clear that $Z_{eq}(j\omega)$ can be made real at the power-carrier frequency $f_o = \omega_o/2\pi$ if

$$\frac{1}{\sqrt{L_2C_2}} = \omega_o \tag{3.58}$$

Next, we define the quality factors at ω_o as

$$Q_1 = \frac{\omega_o L_1}{R_1}, \quad Q_2 = \frac{\omega_o L_2}{R_2}, \quad Q_S = \frac{\omega_o L_2}{R_L + R_2} \tag{3.59}$$

Then,

$$Z_{eq}(j\omega_o) = \frac{\omega_o^2 M^2}{R_L + R_2} = \frac{k^2}{n^2} Q_s{}^2 (R_L + R_2) \tag{3.60}$$

From Eq. (3.56) we have

$$I_2(j\omega_o) = \frac{j\omega_o M}{R_L + R_2} I_1(j\omega_o) \tag{3.61}$$

By substituting Eq. (3.61) into Eq. (3.55) we have

$$V_S = \left(R_1 + \frac{1}{j\omega_o C_1} + j\omega_o L_1 + \frac{\omega_o^2 M^2}{R_L + R_2}\right) I_1(j\omega_o) \tag{3.62}$$

Similarly, $I_1(j\omega_o)$ can be made real at ω_o if

$$\frac{1}{\sqrt{L_1C_1}} = \omega_o \tag{3.63}$$

and

$$I_1(j\omega_o) = \frac{V_S}{R_1 + \frac{\omega_o^2 M^2}{R_L + R_2}} \tag{3.64}$$

It is worth noting that by setting $1/(L_1C_1)^{½} = 1/(L_2C_2)^{½} = \omega_o$, the reflected impedance of the secondary Z_{eq} is purely real, and so is I_1, but I_2 is purely imaginary.

Next, the link voltage gain A_T can be computed by using Eqs. (3.61) and (3.62), and

$$A_T = \frac{|V_o|}{V_S} = \frac{\omega_o k\sqrt{L_1L_2}R_L}{R_L + R_2}\frac{1}{R_1 + \frac{\omega_o^2 k^2 L_1 L_2}{R_L + R_2}} \tag{3.65}$$

The link voltage gain can then be rewritten as

$$A_T = \frac{knQ_1}{(1 + k^2 Q_1 Q_S)}\frac{R_L}{R_L + R_2} \tag{3.66}$$

To compute the link efficiency η_T we note that the load power is $P_o = ½|I_o|^2 R_L$, and the complex source power is $P_{ST} = ½V_S I_S^*$. A simple way to compute η_T is again to make use of $1/(L_1C_1)^{½} = 1/(L_2C_2)^{½} = \omega_o$ and use Eq. (3.61) as follows:

$$P_{ST} = \frac{V_S I_S^*}{2} = \frac{1}{2}(R_1 I_1 - j\omega_o M I_2) I_1^* \tag{3.67}$$

$$\Rightarrow \quad P_{ST} = \frac{1}{2}\left(R_1 \frac{R_L + R_2}{j\omega_o M} - j\omega_o M\right) I_o \frac{R_L + R_2}{-j\omega_o M} I_o^* \tag{3.68}$$

$$\Rightarrow \quad P_{ST} = P_S = \frac{|I_o|^2}{2}\left(R_L + R_2 + R_1 \frac{(R_L + R_2)^2}{\omega_o^2 M^2}\right) \tag{3.69}$$

Hence, the link efficiency is given by

$$\eta_T = \frac{P_o}{P_S} = \frac{1}{1 + \frac{n^2}{k^2}\frac{1}{Q_S{}^2}\frac{R_1}{R_L} + \frac{R_2}{R_L}} \tag{3.70}$$

By comparing Eq. (3.70) with Eq. (3.50), it is clear that the resonant circuit gives a better link efficiency.

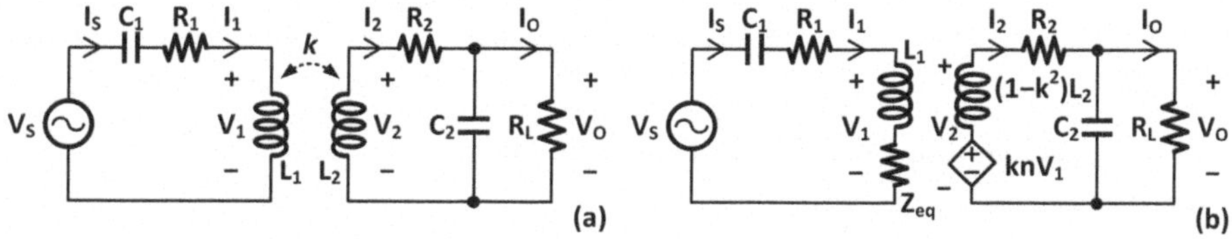

Fig. 3.8 (**a**) The circuit of series-parallel resonant coupled-coils with parasitic resistances, and (**b**) its reflected impedance model

3.3.2 Series-Parallel Resonant Coupled-Coils

If a capacitor is added in series with the primary-coil and a capacitor is added in parallel to the secondary-coil, then we have a pair of series-parallel resonant coupled-coils (S-P coils), as shown in Fig. 3.8.

With reference to Fig. 3.8, the s-domain equations are

$$V_S = (1/sC_1 + R_1 + sL_1)I_1 - sMI_2 \tag{3.71}$$

$$V_2 = sMI_1 - sL_2I_2 = (R_2 + 1/sC_2||R_L)I_2 \tag{3.72}$$

$$I_o = \frac{1/sC_2}{1/sC_2 + R_L}I_2 = \frac{1}{1 + sC_2R_L}I_2 \tag{3.73}$$

The equivalent impedance is given by

$$Z_{eq}(s) = \frac{\omega^2M^2}{sL_2 + R_2 + 1/sC_2 \mid\mid R_L} \tag{3.74}$$

$$\Rightarrow \quad Z_{eq}(j\omega) = \frac{\omega^2M^2(1 + j\omega C_2R_L)}{R_2 + R_L(1 - \omega^2L_2C_2) + j\omega(L_2 + C_2R_2R_L)} \tag{3.75}$$

In [3], $Z_{eq}(j\omega_r)$ is designed to be real by assigning the resonance frequency ω_r to be

$$\omega_r = \sqrt{\frac{1}{L_2C_2} - \frac{1}{C_2^2R_L^2}} = \frac{1}{\sqrt{L_2C_2}}\sqrt{1 - \frac{L_2}{C_2R_L^2}} \tag{3.76}$$

The only advantage of this assignment is that $Z_{eq}(j\omega_r)$ is real, but the disadvantages are at least threefolds. First, ω_r is load-dependent, and if the load (current or resistance) changes, $Z_{eq}(j\omega_r)$ will not be real anymore. Second, it will become apparent later that the parallel-resonant secondary would be good for low-power applications, and therefore, R_L would be large or even very large, for example, $R_L > 100\ \Omega$ or even larger than 1 kΩ. Hence, the factor $L_2/(C_2R_L^2)$ would be much smaller than 1 and can be ignored. Third, using Eq. (3.76) would make the computation of A_T and η_T unnecessarily complicated. Based on the above reasons, it is advisable to follow the same assignment as Eq. (3.58), which is repeated below:

$$\omega_o = \frac{1}{\sqrt{L_2C_2}} \tag{3.77}$$

The equivalent impedance at ω_o is then equal to

$$Z_{eq}(j\omega_o) = \frac{\omega_o^2 M^2 (R_2 + \omega_o C_2 R_L(\omega_o L_2 + \omega_o C_2 R_2 R_L) - j\omega_o L_2)}{R_2^2 + (\omega_o L_2 + \omega_o C_2 R_2 R_L)^2} \tag{3.78}$$

To further proceed with the computation good approximations have to be instituted. Here, we assume that the output voltage is adequately filtered, which requires that

$$\omega_o C_2 R_L >> 1 \tag{3.79}$$

Moreover, the load resistor R_L is much larger than the parasitic resistor R_2, that is,

$$R_L >> R_2 \tag{3.80}$$

Finally, to further simplify the expression we note that $\omega_o L_2 = 1/(\omega_o C_2)$, and we have

$$\frac{L_2}{C_2 R_2 R_L} = \frac{(\omega_o L_2)^2}{R_2 R_L} = Q_2 Q_L \tag{3.81}$$

where the quality factors are defined as follows:

$$Q_1 = \frac{\omega_o L_1}{R_1}, \quad Q_2 = \frac{\omega_o L_2}{R_2}, \quad Q_L = \frac{\omega_o L_2}{R_L} \tag{3.82}$$

From the above assumptions the imaginary part is then negligible compared to the real part, and assigning $\omega_o = 1/(L_2C_2)^{½}$ is a good choice. Next, the real part of the reflected impedance is then given by

$$R_{eq} = \mathrm{Re}(Z_{eq}(j\omega_o)) \approx \frac{\omega_o^2 M^2 \times \omega_o^2 C_2 R_L \times (1 + Q_2 Q_L) C_2 R_2 R_L}{(1 + Q_2 Q_L)^2 (\omega_o C_2 R_2 R_L)^2} \tag{3.83}$$

$$\Rightarrow \quad R_{eq} \approx \frac{\omega_o^2 M^2}{(1 + Q_2 Q_L) R_2} \tag{3.84}$$

Note that the equivalent resistance is a weak function of R_L (the dependence is only through the parameter $(1 + Q_2Q_L)$). In fact, the reason is very simple: at ω_o, $R_L >> |1/j\omega_o C_2|$, and the impedance of the parallel branch is approximated by a small real part plus the impedance of the capacitor $1/(j\omega_o C_2)$, which is then cancelled by the impedance of the inductor $j\omega_o L_2$, leaving only the resistor R_2 plus the small real part, which is together given by $(1 + Q_2 Q_L)R_2$.

The computation of the link voltage gain involves the primary coil, and similar to the secondary coil, eliminating the imaginary part of the primary coil is load-dependent, and we again adopt the simple assignment of

$$\frac{1}{\sqrt{L_1 C_1}} = \omega_o \tag{3.85}$$

Now, Eq. (3.71) is equal to

$$V_S = R_1 I_1(j\omega_o) - j\omega_o M I_2(j\omega_o) \tag{3.86}$$

From Eq. (3.72) we have

$$I_1(s) = \frac{sL_2 + R_2 + 1/sC_2 \,||\, R_L}{sM} I_2(s) \tag{3.87}$$

$$\Rightarrow \quad I_1(j\omega_o) = \frac{R_2 + j\omega_o(L_2 + C_2 R_2 R_L)}{j\omega_o M} I_o(j\omega_o) \tag{3.88}$$

Substituting Eq. (3.88) into Eq. (3.86) gives

$$V_S = \left(R_1 \frac{R_2 + j\omega_o(L_2 + C_2 R_2 R_L)}{j\omega_o M} - j\omega_o M(1 + j\omega_o C_2 R_L)\right) I_o(j\omega_o) \tag{3.89}$$

The link voltage gain $A_T = |I_o| R_L / V_S$ is

$$A_T = \frac{|\, j\omega_o M R_L \,|}{|\, R_1 R_2 + j\omega_o(L_2 + C_2 R_2 R_L) R_1 + \omega_o^2 M^2 (1 + j\omega_o C_2 R_L) \,|} \tag{3.90}$$

With $M = k(L_1 L_2)^{1/2}$ we have

$$A_T = \frac{knQ_1 R_L / R_2}{\left| 1 + \frac{\omega_o^2 M^2}{R_1 R_2} + \frac{j\omega_o L_2}{R_2} + j\omega_o C_2 R_L \left(1 + \frac{\omega_o^2 M^2}{R_1 R_2}\right) \right|} \tag{3.91}$$

Recall $\omega_o C_2 = 1/\omega_o L_2$, and Eq. (3.91) can be rewritten as

$$A_T = \frac{knQ_1 Q_2 / Q_L}{\left| 1 + k^2 Q_1 Q_2 + jQ_2 + j\frac{1}{Q_L}(1 + k^2 Q_1 Q_2) \right|} \tag{3.92}$$

Note that $Q_L << 1$, and we have

$$A_T \approx \frac{knQ_1Q_2}{\left|j(1 + Q_2Q_L + k^2Q_1Q_2)\right|} \tag{3.93}$$

$$\Rightarrow \qquad A_T \approx \frac{knQ_1Q_2}{1 + Q_2Q_L + k^2Q_1Q_2} \tag{3.94}$$

The final task is to compute the link efficiency, with the load power $P_o = ½|I_o|^2 R_L$ and the complex source power $P_{ST} = ½V_SI_1^*$. Hence, using Eq. (3.73) we have

$$P_{ST} = \left(R_1\frac{R_2 + j\omega_o(L_2 + C_2R_2R_L)}{j\omega_oM} - j\omega_oM(1 + j\omega_oC_2R_L)\right)\frac{I_o(j\omega_o)}{2} \times \frac{R_2 - j\omega_o(L_2 + C_2R_2R_L)}{-j\omega_oM}I_o(j\omega_o)^* \tag{3.95}$$

Now, the real part of P_{ST} is $P_S = \mathrm{Re}(P_{ST})$, and

$$P_S = \left(R_1\frac{R_2^2 + \omega_o^2(L_2 + C_2R_2R_L)^2}{\omega_o^2k^2L_1L_2} + R_2 + R_L + \omega_o^2C_2^2R_L^2R_2\right)\frac{|I_o|^2}{2} \tag{3.96}$$

By using the relations in Eqs. (3.79), (3.80) and (3.81), we have

$$P_S \approx \left(R_L + \frac{R_L^2}{\omega_o^2L_2^2}R_2 + \frac{n^2}{k^2}\frac{(1 + Q_2Q_L)^2(C_2R_2R_L)^2}{L_2^2}R_1\right)\frac{|I_o|^2}{2} \tag{3.97}$$

Finally, the link efficiency is given by

$$\eta_T = \frac{½|I_o|^2R_L}{P_S} \approx \frac{1}{1 + \frac{n^2}{k^2}\frac{(1+Q_2Q_L)^2}{Q_2^2Q_L^2}\frac{R_1}{R_L} + \frac{1}{Q_L^2}\frac{R_2}{R_L}} \tag{3.98}$$

3.4 Summary

In this chapter, a design-oriented analysis is used to analyze various coupled-coils driving a resistive load. The configurations include the ideal coupled-coils, the coupled-coils with parasitic resistors, the coupled-coils with series-resonant primary and series-resonant secondary, and the coupled-coils with series-resonant primary and parallel-resonant secondary. For all the configurations, the reflected (or equivalent) impedance, the link voltage gain and the link efficiency are computed. All the expressions are written in a form that directly relates to the circuit parameters such as k, n, R_1, R_2 and R_L, and hence are useful for the design and optimization of the coupled-coils.

References

1. Ko WH, Liang SP, Fung CDF (1977) Design of radio-frequency powered coils for implant instruments. Med Biol Eng Comput 15:634–640
2. Fuller JW (1968) Apparatus for efficient power transfer through a tissue barrier. IEEE Trans Biomed Eng:63–65
3. van Schuylenbergh K, Puers R (2009) Inductive powering. Springer, Dordrecht
4. Irwin JD, Kerns DV Jr (1995) Introduction to electrical engineering. Prentice Hall, New York
5. Hart DW (2011) Power electronics. McGraw Hill, New York

Chapter 4
Circuit Design of CMOS Rectifiers

Abstract This chapter discusses the design considerations and the state-of-the-art designs of CMOS rectifiers for wireless power receivers. First, the forward current conduction capability and the reverse leakage current of on-chip passive diodes are investigated and compared. Next, comparator-based active rectifiers for near-field WPT solutions are discussed and demonstrated with a couple of design examples. Moreover, a near-optimal adaptive on- and off-delay compensation technique is introduced. Other rectifier topologies, such as the delay-locked-loop (DLL) based rectifier and the threshold-voltage compensated rectifier, are also discussed. In addition, rectifiers for RF energy harvesting for far-field WPT are discussed as well.

Keywords Wireless power transfer • AC-DC conversion • Rectifier • Comparator • Delay • RF-DC • Energy harvesting

4.1 Introduction

Wireless power transfer (WPT) has a wide range of applications (from low to high power levels) including radio frequency identification (RFID), implantable medical devices (IMDs), and wireless chargers for portable and wearable devices, and for electric vehicles (EVs). As mentioned in Chap. 1, at different power levels, the considerations on circuit design and device selection can be very different. The focus of this book is on the designs of low to medium power level (<10 W) WPT systems that can be realized in standard CMOS technologies.

The performance of the WPT receiver limits the output power of the whole system, and many works have been focused on the WPT receiver design. Note that the rectifier is a critical part in the receiver chip. For IMD applications, the form factor of the power receiver is one of the principal concerns because these devices have to be as non-invasive as possible. For output power level in the milli-watt range, full on-chip integration is realizable and thus favorable with the elimination of discrete components [1]. For consumer electronics that cater for portable or wearable devices, for example, wireless chargers for mobile phones or tablet computers, the power level is around 10 W. For EV power management, fast-charging the automotive battery system needs very high power in the kilo-watt range, and discrete diodes/transistors fabricated by dedicated processes such as the

Y. Lu, W.-H. Ki, *CMOS Integrated Circuit Design for Wireless Power Transfer*,
Analog Circuits and Signal Processing, DOI 10.1007/978-981-10-2615-7_4

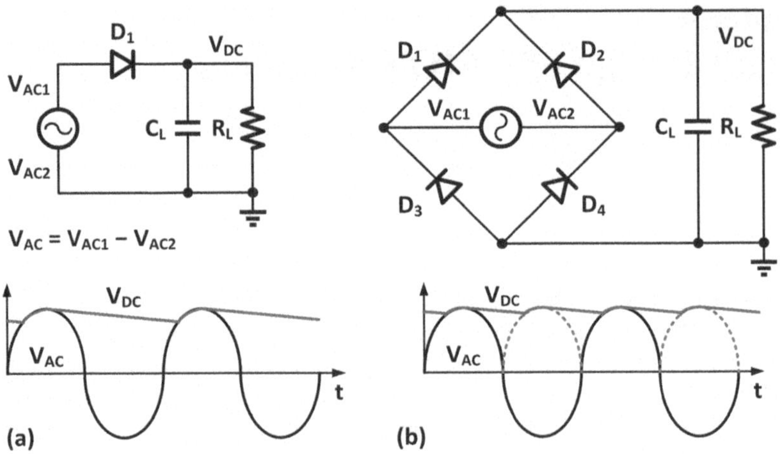

Fig. 4.1 Schematics of (**a**) the half-wave rectifier and (**b**) the full-wave rectifier

III-V direct bandgap Gallium Nitride (GaN) or the laterally diffused MOSFET (LDMOS) processes are commonly employed. Hence, circuit designers are facing the challenges of designing wireless power transfer systems that could range from milli-watt to kilo-watt. For applications of 10 W or lower, the rectifier is preferred to be fully integrated on-chip to avoid using discrete Schottky diodes, and thus saves the circuit board area. For high-power and/or high-voltage applications, GaN devices are popular to provide higher conversion efficiencies, and LDMOS transistors are essential for high-voltage operation.

In WPT receiver and RF energy-harvesting systems, the input AC sources have to be converted into DC sources to power up downstream electronics, which necessitates the process of rectification. Diodes, the simplest semiconductor devices that only allow the current to flow in one direction and block it from flowing in the reverse direction, are commonly used to convert AC power into DC power as shown in Fig. 4.1. Figure 4.1a shows a half-wave rectifier, with one terminal of the AC source (V_{AC2}) tied to ground. For the half cycle when V_{AC1} is higher than V_{AC2}, the diode D_1 will conduct when V_{AC1} goes higher than the DC output voltage V_{DC}; and for the half cycle when V_{AC1} is lower than V_{AC2}, D_1 is reverse-biased for sure. Hence, the half-wave rectifier delivers current from the AC source to the DC output V_{DC} only once in every cycle. Figure 4.1b shows a full-wave rectifier. First, consider the half-cycle when V_{AC1} is higher than V_{AC2}. The diodes D_1 and D_4 will conduct when V_{AC1}–V_{AC2} is higher than V_{DC}. Next, consider the half-cycle when V_{AC1} is lower than V_{AC2}, and D_2 and D_3 will conduct when V_{AC2}–V_{AC1} is higher than V_{DC}. Hence, the full-wave rectifier delivers current to the output at both the positive and the negative peaks of $V_{AC} = V_{AC1} - V_{AC2}$. This means that the time interval for charge transfer of a full-wave rectifier is only half of that of a half-wave rectifier. Therefore, in delivering the same load currents with the same value of output capacitors, the output voltage ripple of a half-wave rectifier is basically two times that of a full-wave rectifier. In addition, the peak amplitude of the pulse current conducted through a half-wave rectifier is roughly doubled compared to that

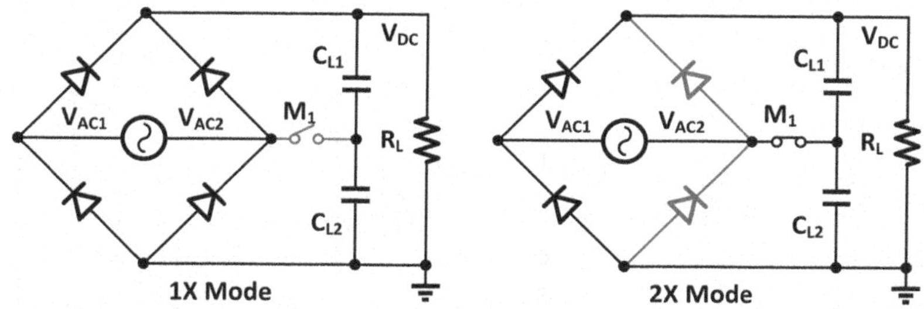

Fig. 4.2 A reconfigurable rectifier (universal rectifier) with 1X (rectifier) mode and 2X (voltage doubler) mode

of a full-wave rectifier, while higher peak current may result in higher conduction loss from the diode.

Ideally, the voltage difference between the two terminals $V_{AC} = V_{AC1} - V_{AC2}$ is a sinusoidal wave; but it could be a distorted sinusoidal wave in practice, which will be discussed later in this chapter.

Figure 4.2 shows a passive reconfigurable rectifier (universal rectifier) with 1X (rectifier) mode and 2X (voltage doubler) mode. It is commonly installed in electrical appliances that are designed to work in different countries with AC supply voltages of either 220 V_{RMS} or 110 V_{RMS}. The reconfigurable rectifier also found its use in WPT applications. To cater for coupling coefficient k variations with distance and/or orientation of the coupled-coils, a reconfigurable rectifier may be employed to increase V_{DC} without increasing the transmitted power. Examples can be found from the loosely coupled inductive power link with active rectifiers to extend the coupling range [2, 3]. It may also be used to modulate the input impedance of the WPT receiver, as in the ideal cases, the input impedance of the 1X mode is four times that of the 2X mode when driving the same resistive load. It is because the 2X mode generates twice the output voltage, and with a fixed load resistor, the power consumed is four times that of the 1X mode, and hence the current from V_{AC} is four times that of the 1X mode. This change in impedance should also be taken into consideration when designing the matching network.

This chapter starts with introducing the basic rectifier topologies and selecting diodes and diode-connected transistors. Emphases are focused on active rectifier designs that are low-cost high-efficiency solutions for various low-power applications. Rectifiers for RF energy harvesting are discussed as well.

4.2 Diodes and Diode-Connected Transistors

Figure 4.3 shows the diodes and the diode-connected transistors that are available in standard CMOS processes. The P+/N-well junction diode is readily available in CMOS processes, because the N-well can be connected to any potential higher than the ground potential. However, the parasitic PN junction (N-well to substrate) may affect high-frequency performance of the circuit and may also introduce DC leakage current.

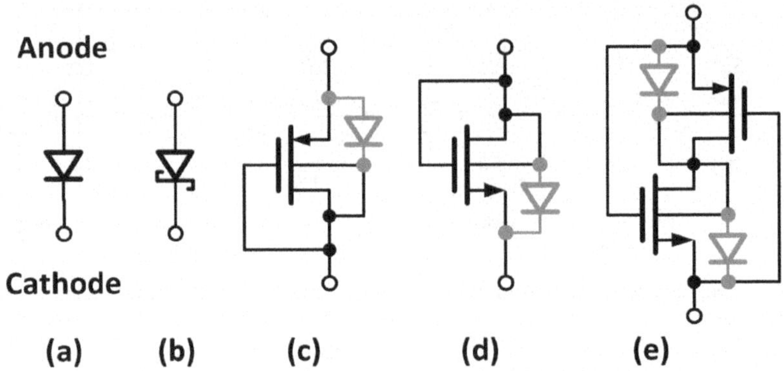

Fig. 4.3 Symbols and schematics of (**a**) normal PN junction diode, (**b**) Schottky diode, (**c**) diode-connected PMOS, (**d**) diode-connected NMOS, and (**e**) composite CMOS diode

A Schottky diode, as shown in Fig. 4.3b, is formed by the junction between a semiconductor and a metal. It has low forward voltage drop and achieves fast switching. Schottky diodes have been widely used in conventional discrete-type AC-DC converters for rectification, and are still in use for various low-voltage high-power applications. Although the process of making Schottky diodes is compatible with CMOS processes, additional masks are needed that increase the manufacturing cost.

Alternatively, PN junction diodes may be replaced by diode-connected transistors as shown in Fig. 4.3c and 4.3d. A PMOS diode-connected transistor is called a PMOS diode, and an NMOS diode-connected transistor is called an NMOS diode. They could be used to implement a cost-effective passive rectifier. It should be noted in rectifier designs the body bias of the diode-connected MOSFET should be connected to its drain terminal as shown in the figure, to make sure that the parasitic diode has the same polarity as the MOS diode; otherwise, the parasitic diode will be forward biased when a reverse voltage is applied, and large reverse current will flow. This body bias connection is different from that in a usual analog circuit.

Performance comparison of different diodes can be evaluated by their on/off current ratios; and to compare the turn-off scenarios of different diodes, we need to study the I-V characteristic of MOS transistors working in the sub-threshold region. For an NMOS transistor, the sub-threshold I-V characteristic is given by

$$I_{\mathrm{D}} = I_0\left(e^{\frac{V_{\mathrm{GS}}}{\zeta V_{\mathrm{T}}}} - 1\right), \tag{4.1}$$

where $\zeta > 1$ is the non-ideality factor, V_{GS} is the gate-to-source voltage, and $V_{\mathrm{T}} = kT/q$ is the thermal voltage. The non-ideality factor ζ for a PN junction diode is usually lower than that of a MOS transistor, and therefore results in a larger exponential $I_{\mathrm{D}}/V_{\mathrm{D}}$ relationship. With a typical value of ζ, V_{GS} has to decrease by roughly 80 mV for I_D to decrease by one decade at room temperature [4]. If low-threshold voltage transistors are used to give a lower V_{GS}, the reverse leakage current would increase exponentially.

Consider the case when the wirelessly transmitted power is discontinued, the energy stored in the devices will be drained by leakage currents and thus the stand-by time would be reduced. The reverse leakage current of an NMOS diode is given by

$$I_D = I_0 e^{\frac{-|V_{GS}|+V_{TH}}{\zeta V_T}}\left(1 - e^{\frac{-|V_{DS}|}{V_T}}\right), \tag{4.2}$$

Where V_{TN} is the threshold voltage and V_{DS} is the drain-to-source voltage. Obviously, the reverse leakage current has an exponential relationship with two parameters: V_{GS} and V_{DS}. In a diode-connect MOSFET, $V_{GS} = V_{DS}$. If short-channel transistors in an advanced process are employed to increase the conducting current capability and to reduce the parasitic components, the reverse leakage current will also be increased, because the shorter channels are relatively easier to be depleted with a lower $|V_{DS}|$, and the charge carriers will then shoot through the channel. Since the MOS diode and its parasitic junction diode are in parallel, the reverse leakage current of a MOS diode consists of two parts: the leakage current through the channel, and the leakage current through the parasitic junction diode. The second part is much smaller than the first part because the junction area is small and the channel is short.

Figure 4.4a shows the simulated I-V curve of a diode-connected 1.8 V PMOS diode with ten fingers of $W = 10$ μm and $L = 0.2$ μm. When the reverse bias is small, the I-V curve follows Eq. (4.2). When the reverse bias is lower than −2 V, in this case, the channel starts to be depleted, and the reverse current starts to increase significantly. In the forward conduction region, the forward current of the PMOS diode increases significantly when the parasitic junction diode is turned on at $V_D > 0.8$ V.

To reduce the reverse leakage current of a MOS diode, a composite CMOS diode as shown in Fig. 4.3e was proposed in [5], and has been applied in low-power CMOS rectifier designs [6, 7]. The simulated I-V curve of the composite CMOS diode is shown in both Fig. 4.4a and 4.4b and compared with other diodes. In the forward-bias mode, the composite CMOS diode can be regarded as two serially connected forward-bias NMOS and PMOS diodes. The forward current is thus comparable to that of the PN junction diode (on the same order of magnitude). When the composite CMOS diode is reversely biased, both transistors operate with negative $|V_{GS}|$ in the accumulation region. Starting from $|V_{DS}| = 0$ V, the reverse current increases initially as V_{DS} goes negative, due to the factor $(1 - e^{-|VDS|/VT})$. However, as the reverse bias goes further negative, the reverse current is then dominated by the $e^{-|VGS|/\zeta VT}$ term, and decreases with a slope of less than 200 mV per decade. Note that for both the PMOS and the NMOS transistors, their $|V_{DS}|$ are roughly half of their $|V_{GS}|$ because

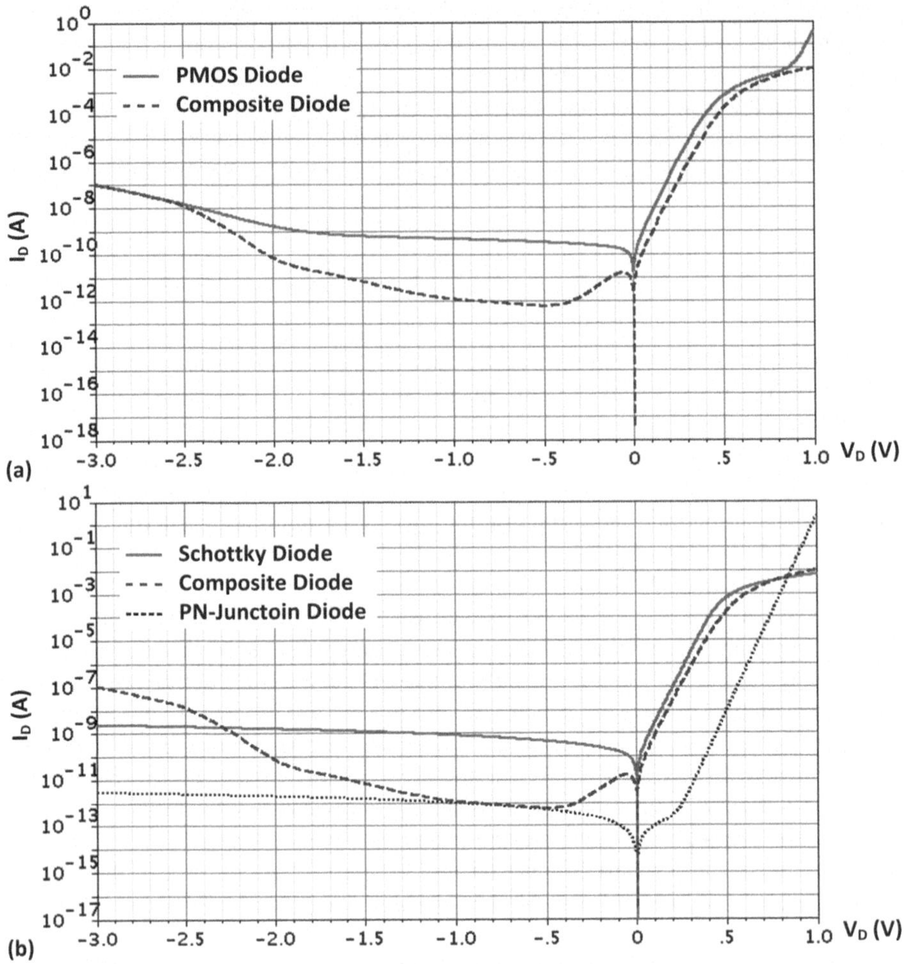

Fig. 4.4 Comparisons of the I-V characteristics (**a**) between PMOS diode and composite CMOS diode, and (**b**) between Schottky diode, composite CMOS diode, and normal PN junction diode

$$|V_{DSN}| + |V_{DSP}| = |V_{GSN}| = |V_{GSP}|. \tag{4.3}$$

This is different from the case of the simple PMOS and NMOS diodes, where V_{DS} is always equal to V_{GS}. As can be explained by Eq. (4.2), this leads to a low leakage solution by limiting one of the exponential factors ($e^{-|VDS|/VT}$). When a reverse-bias voltage of some hundreds of milli-Volts is applied, the reverse leakage current of a composite CMOS diode is as low as the junction leakage current.

4.3 Comparator-Based Active Rectifiers

For applications with power level higher than mW, using diode-connected transistors may not achieve high efficiency, because their $|V_{GS}|$ can only be slightly higher, say by 100 mV, than their $|V_{TH}|$, and result in large chip area. Instead, active diodes that are essentially comparator-controlled MOSFETs can be used to replace the passive diodes in a passive rectifier, and to form an active rectifier, as shown in Fig. 4.5. Active diodes have low forward voltage and thus low loss, and are particularly important for low input voltage applications. In addition, a fully-integrated active rectifier reduces the number of discrete components needed and thus reduces the system cost.

Two important benchmark parameters in evaluating active rectifiers are the voltage conversion ratio M and the power conversion efficiency PCE. The voltage conversion ratio is defined as

$$M = \frac{V_{DC}}{|V_{AC}|}, \tag{4.4}$$

where $|V_{AC}|$ is the amplitude of the input AC signal to the rectifier, and V_{DC} is the averaged rectified output DC voltage. The power conversion efficiency of an AC-DC converter is defined as

$$\mathrm{PCE} = \frac{P_{OUT}}{P_{IN}} = \frac{V_{DC}^2/R_L}{\frac{1}{N\cdot T}\int_{t_0}^{t_0+N\cdot T} V_{AC}(t)\cdot I_{AC}(t)\mathrm{d}t}, \tag{4.5}$$

where T is the period of the input sinusoidal wave, N is the number of cycles that are integrated for P_{IN} calculation, and $V_{AC}(t)$ and $I_{AC}(t)$ are the instantaneous voltage and current of the AC source.

For the passive rectifiers, as discussed in Sect. 4.2, both the voltage conversion ratio and the power conversion efficiency are low because the forward voltage drop V_D of a PN junction diode is around 0.6–0.7 V, and that of a Schottky diode is around 0.15–0.45 V. An active rectifier that only uses CMOS transistors is shown in Fig. 4.6. The two high-side diodes in the passive full-wave rectifier that shown in

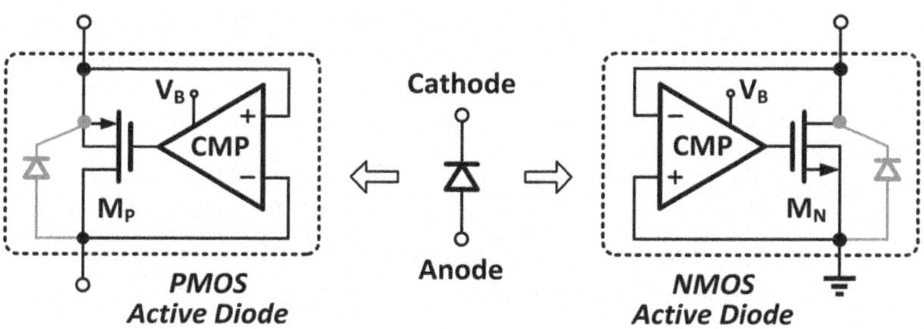

Fig. 4.5 PMOS and NMOS active diodes

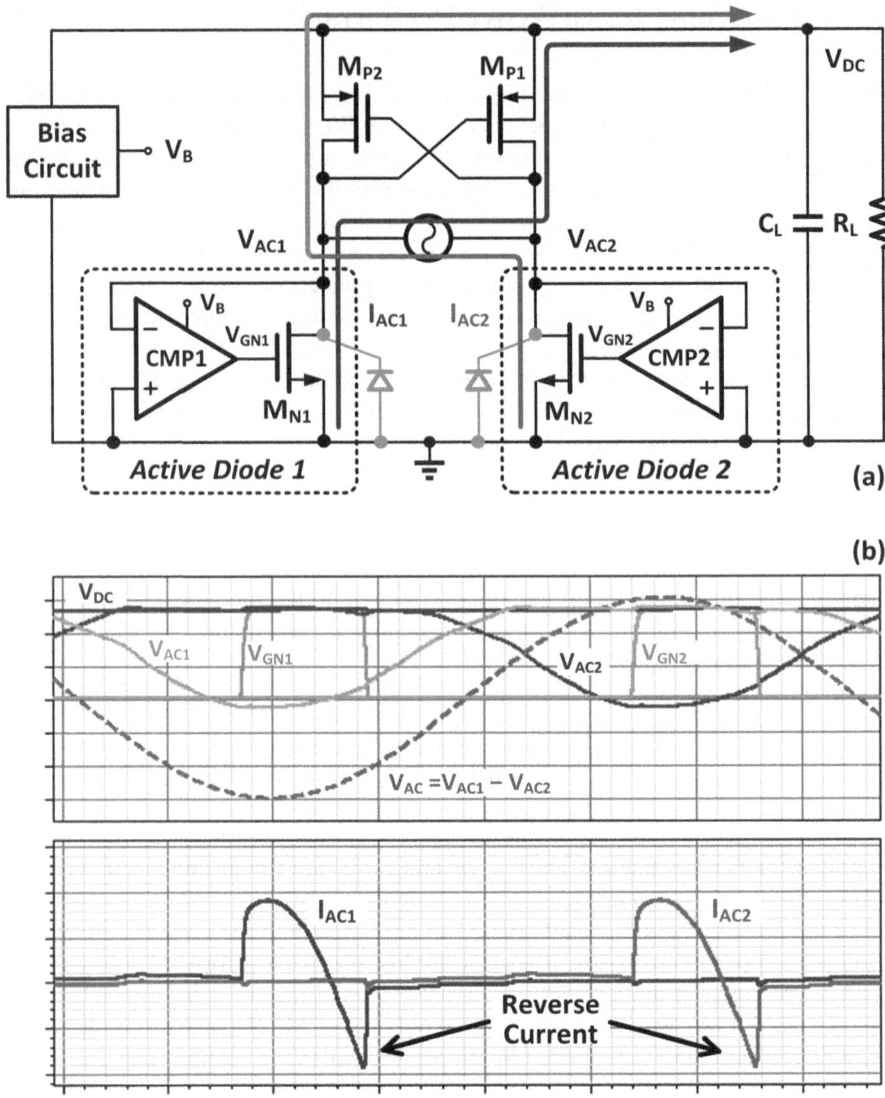

Fig. 4.6 (**a**) An active full-wave rectifier with (**b**) simulated AC voltage and current waveforms showing the reverse current problem in active rectifier

Fig. 4.1b are replaced by two cross-coupled PMOS transistors, and the low-side diodes are replaced by two comparator-controlled NMOS switches (active diodes). In this configuration, the forward voltage drops are reduced from $2V_D$ to $2V_{DS}$, where V_{DS} is the turn-on voltage (that is equal to the drain-to-source voltage) of the power switches.

The operation principle of the active full-wave rectifier can be described as follows. Assume the process starts with V_{AC1} going down and V_{AC2} going up. When

$V_{AC2} - V_{AC1} > |V_{THP}|$ (threshold voltage of $M_{P1,2}$), M_{P1} is turned on and therefore $V_{AC2} = V_{DC}$. Then V_{AC1} swings below the ground voltage, the comparator CMP1 turns on the switch M_{N1}, and I_{AC1} charges up V_{DC} through the AC source. After V_{AC} reaches its low peak point, V_{AC1} starts to go up. When V_{AC1} swings above zero, M_{N1} is then turned off by CMP1, finishing one half cycle of the full-wave rectification period. During the next half of the AC input cycle, the other half of the rectification circuit will conduct in a similar fashion as described above. Thus, by replacing the four diodes of a full-wave rectifier with power transistors, M can be significantly increased especially when the input amplitude is low. However, when operating at a high frequency, such as 13.56 MHz, the comparator delay and the gate-drive buffer delay would affect the efficiency of the rectifier. As demonstrated by the simulated I_{AC} waveforms, the reverse current will occur if the large power switches are not turned off immediately when V_{AC1} or V_{AC2} is higher than the ground voltage.

In this architecture, the main losses include the conduction loss and the switching loss of the power transistors, and the static power loss of the comparator. In terms of power loss and PCE, the extra loss caused by the reverse current is already included in the above mentioned conduction loss, as the energy goes from the DC output back to the AC input is recycled. But in terms of energy extraction from the source, the reverse current obviously reduces the maximum power that can be extracted from the source. From either of the perspectives, the reverse current should be eliminated.

A simple example of a comparator-based active diode is shown in Fig. 4.7 [8]. The two comparators CMP1 and CMP2 are common-gate push-pull differential-input comparators, with the bias currents provided by M_7 and M_8. As described above, when $V_{AC2} - V_{AC1} > |V_{THP}|$, M_{P2} is turned on, and V_{AC1} keeps going down. When V_{AC1} swings below 0 V, M_1 of CMP1 sinks a larger current than M_2 does. As M_4 is diode-connected, therefore, V_2 drops with V_{AC1}, and reduces V_{GS3}. The current of M_1 is mirrored to M_6 through M_5, causing M_6 to source a larger current than M_3 can sink, thus driving V_{OUT} as well as V_{GN1} high and turns on the active-diode switch M_{N1}. In the other half cycle, V_{AC1} goes up and V_{AC2} goes down, then M_{N2} will be turned on in a similar manner and conducts current from ground to V_{OUT} through the AC source.

4.3.1 Delay Compensation Schemes

Due to the speed of the comparator and the gate-drive buffer, the propagation delay of the rising edge (from low to high, t_{pLH}) will shorten the current conduction time Δt, limiting the highest operation frequency of the active rectifier. On the other hand, the propagation delay of the falling edge (from high to low, t_{pHL}) forces the power NMOS transistors $M_{N1,2}$ to be turned off late, and the charge of the output capacitor will flow back to ground through $M_{N1,2}$, resulting in reverse conduction current. Moreover, $M_{N1,2}$ have to be large to handle a large output current,

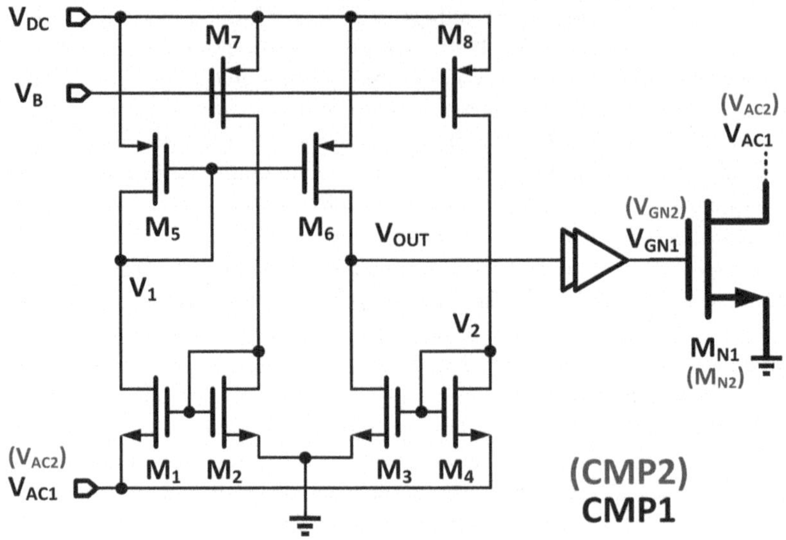

Fig. 4.7 A comparator-based active diode with push-pull common-gate differential input pairs

increasing the response time of the active diodes. This problem is more pronounced when the operation frequency is high (for example, 13.56 MHz) and the input amplitude $|V_{AC}|$ is low (for example, below 1.5 V), because delay time of comparators and buffers are inversely proportional to the supply voltage.

Many schemes have been proposed to compensate for the delays of the active diodes [8–17]. Comparators with unbalanced bias currents or asymmetric differential inputs are used to set an artificial input offset voltage to compensate for the delay and to turn the power switches on and off properly. Prior comparator offset-control schemes for reverse current control fall into one of the cases sketched in Fig. 4.8.

Figure 4.8a shows the circuit symbol of an artificial input offset circuit with an Enable pin. The operation is as follows: when the Enable bit is high, a non-zero offset voltage is added (to the comparator); and when the Enable bit is low, the offset voltage is zero, which means that the symbol represents a shorted wire. To get familiar with the defined symbol, the well-known hysteretic comparator is given in Fig. 4.8b, for showing a complementary example to the delay compensation schemes given in Fig. 4.8c. When the hysteretic comparator just switches to output a "0" (or "1"), a positive offset will be added to its negative (or positive) input terminal, such that the hysteretic comparator output is relatively more difficult to be switched back over to "1" (or "0") due to jitters. For the delay compensation schemes, it is the other way round, as discussed in details below.

Consider the case when V_{AC1} is swinging down initially. In Case 0, the comparator has no added offset, and thus there is reverse current due to delays. In Case 1, an offset voltage of $+V_{OS1}$ is added between V_- of the comparator and V_{AC1}. When V_{AC1} swings down to 0 V, V_- is still at $+V_{OS1}$, and it takes some time before

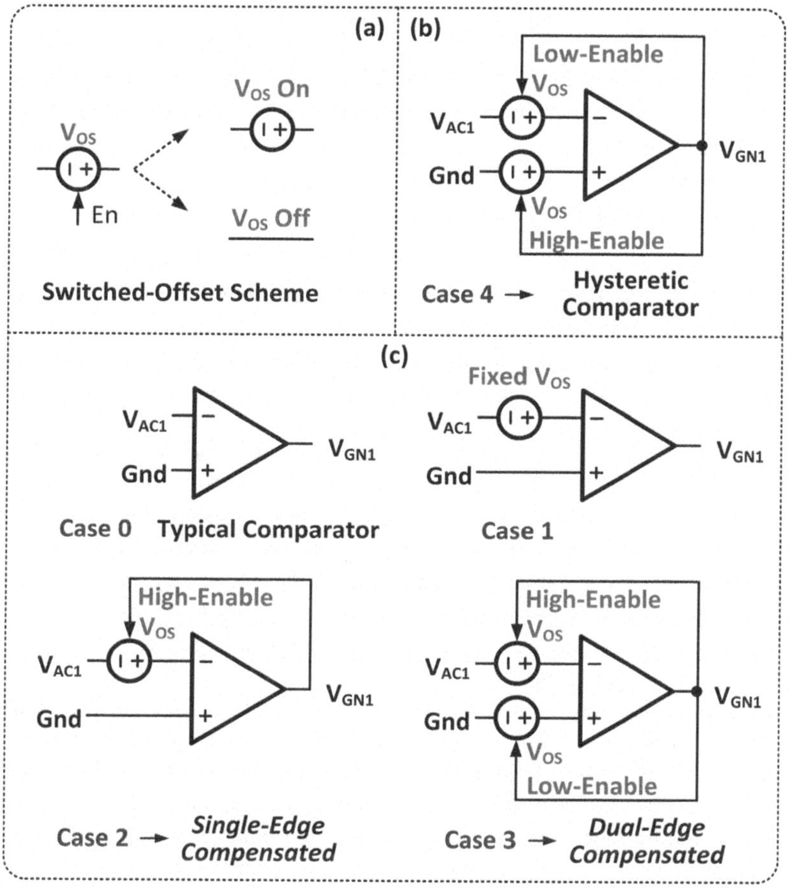

Fig. 4.8 (**a**) Symbol of the switched-offset circuit for comparator delay compensation, and schematics of (**b**) the hysteretic comparator and (**c**) comparator offset-control schemes for reverse current control

V_- swings to 0 V and V_{GN1} trips. Hence, V_{GN1} is switched high later than without $+V_{OS1}$. After V_{AC1} reaches the lowest voltage, it swings back up. Due to the offset voltage, V_- reaches 0 V earlier than V_{AC1} and hence, V_{GN1} is switched low earlier. If $+V_{OS1}$ is designed correctly, reverse conduction can be prevented; but note that M_{N1} is turned on later than required. In Case 2, $+V_{OS1}$ is added only when V_{GN1} is high, and hence the timing of turning on M_{N1} is the same as if no $+V_{OS1}$ is added. In Case 3, besides adding $+V_{OS1}$ to V_-, a second offset voltage $+V_{OS2}$ is added between V_+ of the comparator and Gnd. When $+V_{OS2}$ is added, V_{GN1} is switched high earlier, and $+V_{OS2}$ is added only when V_{GN1} is low. If the offset voltage levels are designed correctly, both the rising edge delay and the falling edge delay can be compensated.

To evaluate different comparator offset-control schemes on the performance of the corresponding active rectifiers, the parameter Crest Factor (CF) that used in evaluating electrical appliances and also used in signal analyses could be employed. The crest factor is defined as the ratio of the peak load current to its RMS (root mean square) value, that is,

$$\mathrm{CF} = \frac{I_P}{I_{\mathrm{RMS}}}. \tag{4.6}$$

A DC current has the minimum CF of 1. A higher CF means a higher peak current for the same average load current, resulting in a larger I^2R conduction loss. Hence, for a load current profile that has a high CF, larger power transistors are needed to achieve a high voltage conversion ratio. Routing metal wires have to be wider as well. The cresting factors of the above cases are considered below.

The scenarios of current conduction of Case 0 to Case 3 are sketched in Fig. 4.9, and discussed as follows. Case 1 was implemented in [9], with constant offset introduced to the comparators using unbalanced bias currents. The power NMOS switches are turned off earlier by t_{d} to eliminate the reverse current; however, they are turned on later by $2t_{\mathrm{d}}$ than the ideal case, and the conduction time Δt_1 is $2t_{\mathrm{d}}$ shorter. For the same load current, the peak current I_{p1} has to be higher, limiting its operation at a higher frequency. The effects of the delays get worse when the AC input amplitude $|V_{\mathrm{AC}}|$ is low.

In [8], self-biased active diodes were employed, and to reduce or eliminate t_{d}, a reverse current control (RCC) scheme (Case 2) is introduced as shown in Fig. 4.10. M_9 serves as the RCC switch, which is turned on together with the active diode. The bias current of this active diode is determined by two diode-connected transistors M_3 and M_5, and thus both the bias current and the operation of the RCC transistor are highly affected by $|V_{\mathrm{AC}}|$ and process variations, making the rectifier hard to be optimized over a wide input range. Moreover, as the reverse current control is realized by a time-varying offset, the artificial offset would disappear right after the power NMOS transistor is turned off. If the RCC switch turns on prematurely (for example, due to process variations) and turns off the active diode while $|V_{\mathrm{AC}}|$ is still higher than V_{DC}, the comparator will go high again in the same cycle. Simulation waveforms of the described scenario are shown in Fig. 4.11. This multiple-pulsing problem incurs larger switching loss, especially in light-load condition when switching loss dominates, and the efficiency is reduced. The remedy is to use SR latches that allow only one switch-on per cycle [10] and the details are to be discussed in Sect. 4.3.4.

Case 3 was realized in [11], with an offset-control circuit to control the comparators to compensate for both turn-on and turn-off delays such that Δt_3 could be maximized (to achieve the lowest CF). However, it suffers from the same and even worse multiple-pulsing problem as [8], as the dynamic offset (+/− offset voltages) transition of the offset-control circuit is unstable: when the comparator outputs a "1" (or "0"), the dynamic offset flips the output to "0" (or "1"), and this positive feedback makes the comparator undergoes self-oscillation. This phenomenon is

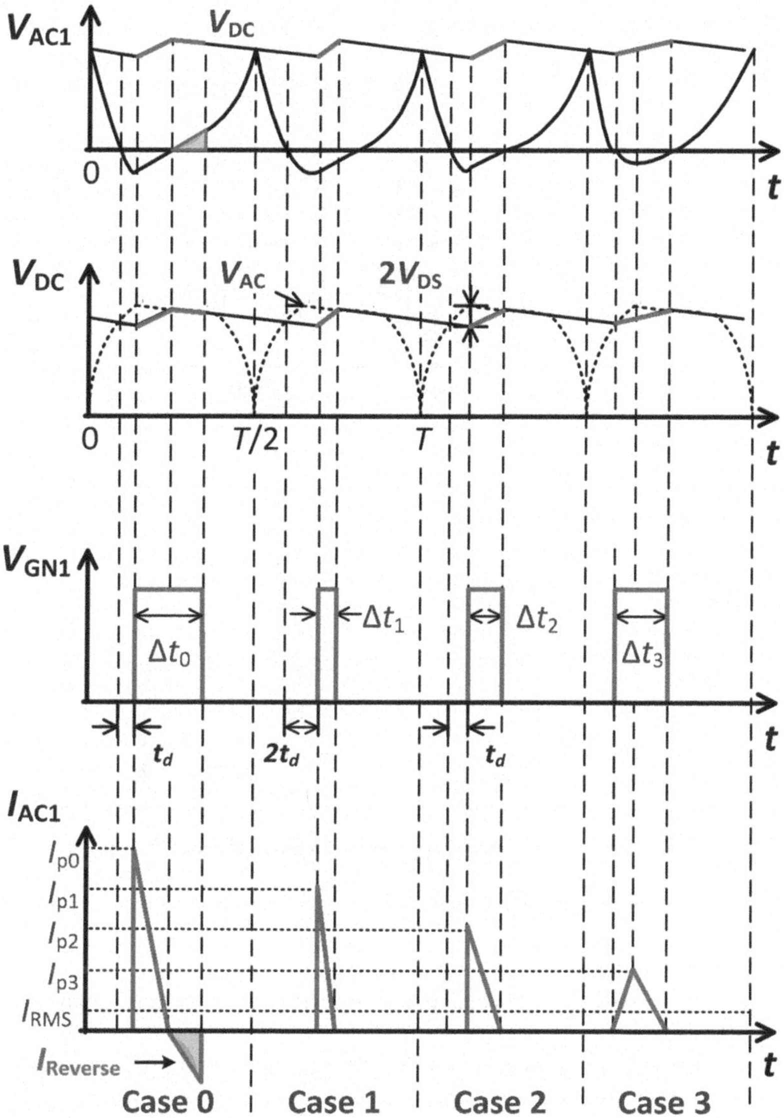

Fig. 4.9 Conceptual waveforms of the voltages and conducted currents for active rectifiers with different comparator delay compensation schemes

what a hysteretic comparator is designed to avoid. To make the scheme work, a delay cell t_{dp} in the offset-control path is added in [11]. If the delay time t_{dp} is large enough (comparable to Δt), the comparator could be stable, but then the power switch would be turned on for at least the duration of t_{dp} that limits the minimum conduction time ($\Delta t_{min} > t_{dp}$). This property makes its operation more like a

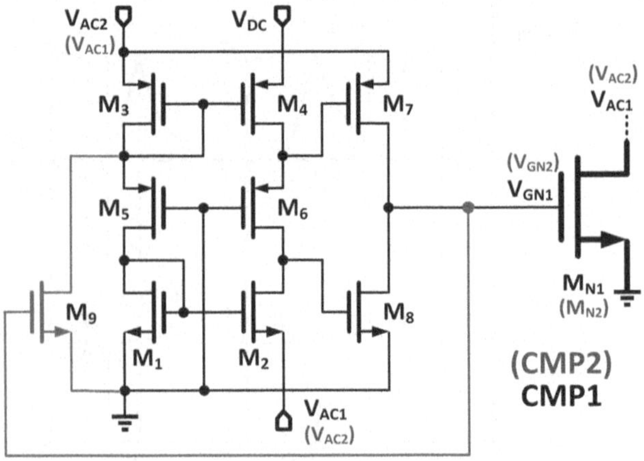

Fig. 4.10 A self-biased active diode with reverse current control

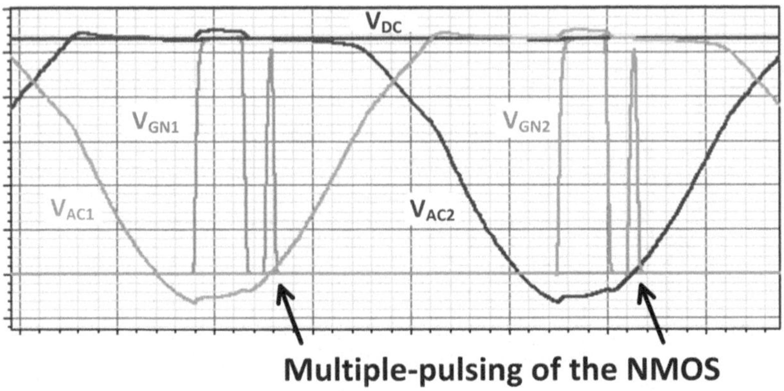

Fig. 4.11 Simulated waveforms of the multiple-pulsing problem associated with dynamic offset schemes

constant on-time control at light-load condition. As a result, its light-load efficiency is degraded.

Recently, Case 3 was also realized in [16] with more comprehensive considerations. Figure 4.12 shows a near-optimum switched-offset compensation scheme implemented for both on- and off-delays with additional sample-and-hold (S&H) voltage-sensing feedback loops. The values of V_{AC1} at both turn-on and turn-off points of the active diodes are sampled on respective sampling capacitors C_{SAMPLE} of the feedback loops. The values of the offset voltages for optimum turn-on/−off timing control are then automatically adjusted by the outputs of the operational transconductance amplifiers (OTAs). As such, the optimum comparator offset will be gradually and automatically adjusted according to the AC input amplitude, WPT frequency, bias current, and load current conditions etc. Furthermore, to avoid the multiple-pulsing problem, a hard one-shot logic that allows the power transistors to turn on only once per cycle is installed. Consequently, zero-voltage switching

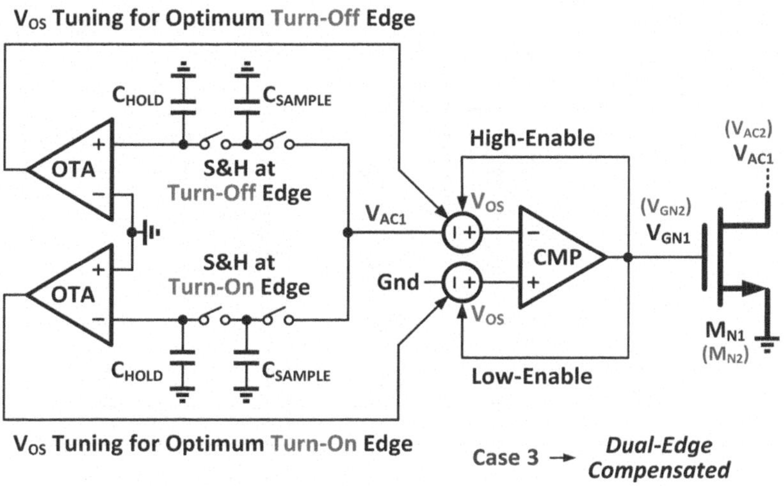

Fig. 4.12 Case 3: Active diode with additional feedback loops for near-optimum offset control

(ZVS) of the power transistors is realized, and the reverse current in [16] is eliminated elegantly.

To summarize, Case 2 has longer Δt than Case 1, and Case 3 has the largest Δt in the ideal situation. However, a comparator with both compensated turn-on and turn-off delays is logically unstable: the hysteresis goes the opposite direction as a normal hysteretic comparator goes, and the robustness of the rectifier is degraded, and special logic blocks are needed to achieve a stable operation. Last but not least, the recent solution implemented in [16, 17] with on–/off-delay sampled control loop solves the reverse current problem with additional circuit blocks.

4.3.2 Delay Time Analysis of Active Diodes

One aspect of active rectifier design that has not yet been discussed is its bias current generation circuit. But before discussing the current source to be used to bias CMP1 and CMP2, let us investigate the delay time of the active diode $t_{d,AD}$ first.

The active-diode delay $t_{d,AD}$ consists of the comparator delay $t_{d,C}$ and the gate-drive buffer delay $t_{d,B}$. To simplify the calculation, as shown in Fig. 4.13, the supply voltage of a typical comparator and buffer is connected to V_{DC} that is AC input dependent, and the input signal is assumed to be sinusoidal, that is, $V_{IN}(t) = \alpha|V_{AC}| \times \sin(2\pi t/T)$, where α is a scaling factor. Let the trip point of the inverter buffer be the 50% point of V_{DC}, that is, $M|V_{AC}|/2$ ($M \approx 0.9$). Consider the comparator output capacitor C_{OUT}. The charging/discharging current $I_{CD}(t)$ of C_{OUT} can be approximated by the small-signal model current as

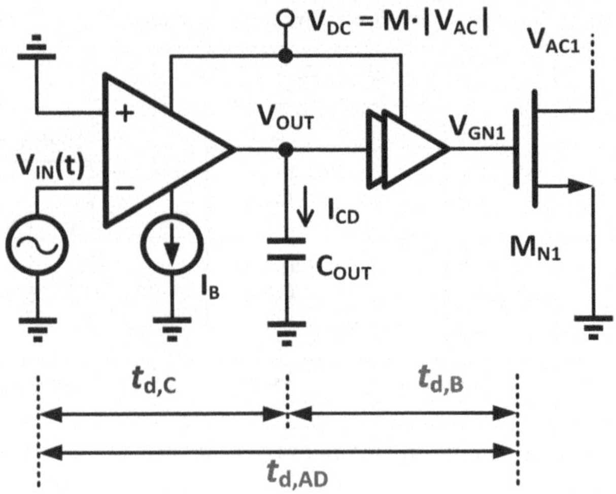

Fig. 4.13 Simplified schematic of active diode 1 with emphasis on supply voltage, input signal and load capacitor for delay calculations

$$I_{\mathrm{CD}}(t) = g_m V_{\mathrm{in}}(t) = g_m \alpha |V_{\mathrm{AC}}| \sin(2\pi t/T), \tag{4.7}$$

where g_m is the transconductance of the comparator input stage. The comparator delay $t_{\mathrm{d,C}}$ is the time that $I_{\mathrm{CD}}(t)$ charges C_{OUT} from 0 V to $V_{\mathrm{DC}}/2$:

$$\begin{aligned}\int_0^{t_{\mathrm{d,c}}} I_{\mathrm{CD}}(t)\mathrm{d}t &= \int_0^{t_{\mathrm{d,c}}} g_m \alpha |V_{\mathrm{AC}}| \sin(2\pi t/T)\mathrm{d}t \\ &= C_{\mathrm{OUT}} V_{\mathrm{DC}}/2 = C_{\mathrm{OUT}} \mathrm{M} |V_{\mathrm{AC}}|/2.\end{aligned} \tag{4.8}$$

For $t << T$, $\sin(2\pi t/T) \approx 2\pi t/T$, and we have

$$\frac{1}{2} g_m \alpha |V_{\mathrm{AC}}| 2\pi t_{\mathrm{d,C}}^2 / T = C_{\mathrm{OUT}} \mathrm{M} |V_{\mathrm{AC}}|/2. \tag{4.9}$$

Therefore,

$$t_{\mathrm{d,C}} = \sqrt{\frac{M C_{\mathrm{OUT}} \mathrm{T}}{2\pi \alpha g_m}} = \sqrt{\frac{M C_{\mathrm{OUT}} \mathrm{T}}{2\pi \alpha \sqrt{2 I_{\mathrm{B}} k_{\mathrm{n}} (W/L)}}}. \tag{4.10}$$

From (4.10), we have $t_{\mathrm{d,C}} \propto \sqrt[4]{1/I_{\mathrm{B}}}$. If I_{B} is independent of V_{DC}, then so does $t_{\mathrm{d,C}}$. The simulated $t_{\mathrm{d,C}}$ of Fig. 4.13 is plotted in Fig. 4.14, which verified the above calculations that $t_{\mathrm{d,C}}$ has a very weak dependence on I_{B}. Assume that the bias current generation circuit would provide an I_{B} that is more or less proportional to V_{DC}, then in such a case,

$$t_{\mathrm{d,C}} \propto \sqrt[4]{1/I_{\mathrm{B}}} \propto \sqrt[4]{1/|V_{\mathrm{AC}}|}. \tag{4.11}$$

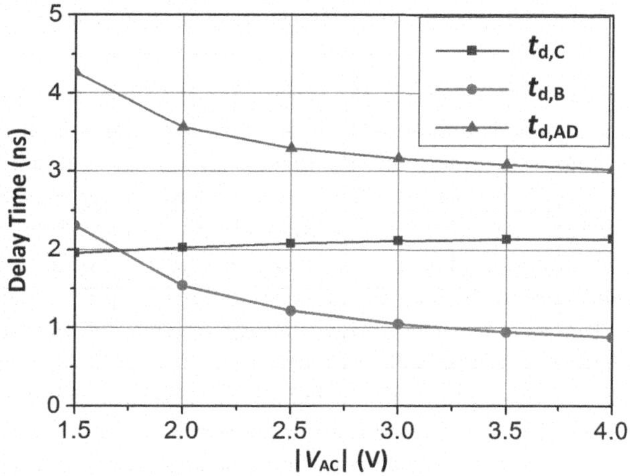

Fig. 4.14 Simulated active diode delay $t_{d,AD}$, comparator delay $t_{d,C}$ and gate drive buffer delay $t_{d,B}$ with respect to $|V_{AC}|$

On the other hand, the buffer delay $t_{d,B}$ would change inversely proportional to the supply voltage, according to the calculated propagation delay t_P (the time to reach the 50% point) of an inverter [18]:

$$\begin{aligned} t_P &= 0.69\frac{3}{4}\frac{C_L V_{DC}}{I_{DS}}\left(1-\frac{7}{9}\lambda V_{DC}\right) \\ &= 0.52\frac{C_L V_{DC}\left(1-\frac{7}{9}\lambda V_{DC}\right)}{k'(W/L)V_{DSAT}(V_{DC}-V_{TH}-V_{DSAT}/2)}, \end{aligned} \tag{4.12}$$

where I_{DS} is the drain current, $k' = \mu C_{ox}$, V_{TH} is the threshold voltage, λ is the channel-length modulation factor, and C_L is the parasitic gate capacitance of the power transistor M_{N1}. Assuming that $V_{DC} >> V_{TH}$, $V_{DSAT} = V_{DC} - V_{TH}$, and $\lambda = 0$, Eq. (4.12) can be written as

$$t_P = t_{d,B} \approx \frac{2C_L}{k'(W/L)V_{DC}}. \tag{4.13}$$

From the above first-order approximation, we learn that $t_{d,B}$ is approximately inversely proportional to the supply voltage. As previously mentioned, delay compensation of the comparator is realized by introducing an input offset voltage such that the power transistor is turned on and off earlier. Similarly, to better compensate for $t_{d,B}$, we need an adaptive current source that is approximately inversely proportional to V_{DC}. Although the first-order model could provide insight into the operation of the circuit to determine the dominant parameters, trial-and-error simulations have to be conducted in practical designs.

4.3.3 Biasing Circuits of Active Rectifiers

For a WPT receiver, the active rectifier has to start working first before the subsequent circuits could work. Therefore, for a robust design, the biasing circuit of the active rectifier should be self-started and not require a start-up circuit. Moreover, a start-up circuit consumes quiescent current even when it finishes its job and is cut open from the main circuit, and thus affects the efficiency. Hence, the commonly used supply-insensitive Widlar current source may not be the best candidate for biasing the active rectifier in a WPT receiver.

Figure 4.15 shows a few biasing circuits without the need of a start-up circuit that were used in WPT receivers [2, 8–16]. In [2, 9], a simple current source that is formed by a diode-connected MOS transistor driving a resistor is used; and in [8, 11, 12], self-biased current sources are used. For the above schemes, the bias current is approximately proportional to the input voltage amplitude $|V_{AC}|$. In [10, 15, 16], the peaking current source (PCS) is used to bias the comparators so that the bias current would stay approximately constant when $|V_{AC}|$ changes.

The bipolar peaking current source was invented in [19], and the CMOS peaking current source was discussed in [20]. The frequency response of the CMOS PCS was compared with the Widlar current source in [21]. With reference to Fig. 4.15c, it is easily shown that I_1 and I_B are given by

$$I_1 = \frac{V_{DD} - V_{GS1}}{R_B} = \frac{1}{2}k_n\left(\frac{W}{L}\right)_1 (V_{GS1} - V_{THN})^2 \tag{4.14}$$

$$I_B = \frac{1}{2}k_n\left(\frac{W}{L}\right)_2 (V_{GS1} - V_{THN} - I_1R_1)^2. \tag{4.15}$$

For I_B to be insensitive to the change of I_1, we set dI_B/dI_1 to zero, and using (4.14) and (4.15), the condition for locating the maxima is

$$I_1R_1 = \frac{1}{2}(V_{GS1} - V_{THN}), \tag{4.16}$$

and the bias current peaks at the nominal V_{DC} that we use to calculate I_1, that is, $I_1 = (V_{DC,nom} - V_{GS1})/R_B$. A simple assignment is to set $(W/L)_2 = 4(W/L)_1$ to give $I_B = I_1$.

However, as previously mentioned, for a constant $t_{d,AD}$ over a wide range of AC input amplitude, the bias current of the active diode should be approximately inversely proportional to $|V_{AC}|$. In the open-loop designs of [3, 14], to further improve the performance at low $|V_{AC}|$ a dual-peaking current source was used, as shown in Fig. 4.16b. By setting the transistor size ratios among M_{N1}, M_{N2} and M_{N3} appropriately, the bias current can be set to be quasi-inversely proportional to $|V_{AC}|$ ($\approx V_{DC}$) as shown in Fig. 4.16c, labeled as a QIPV bias, or can be set to a bias current that is more insensitive to supply variations as shown in Fig. 4.16d. By using the QIPV bias, the comparator offset could be well controlled over the whole AC

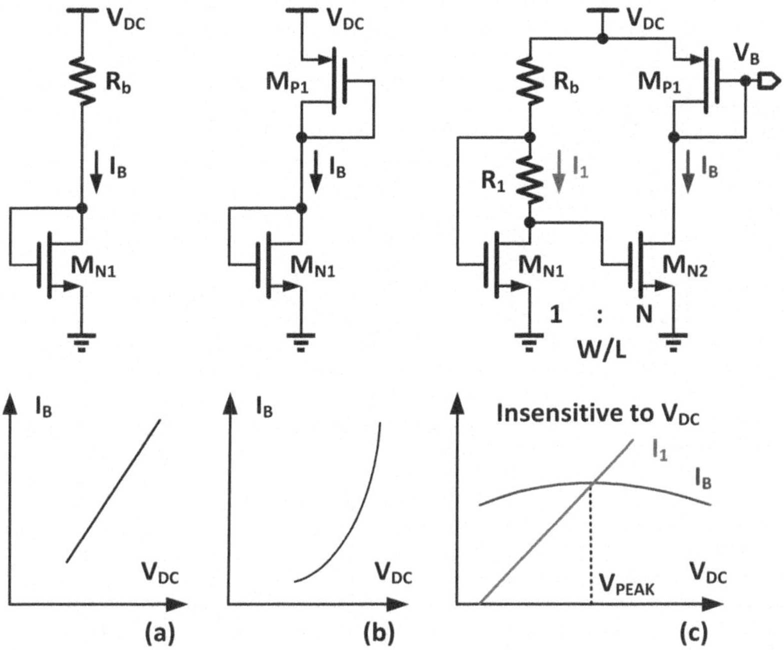

Fig. 4.15 Biasing circuits without start-up requirement: (**a**) resistor-based biasing, (**b**) diode-connencted transistor biasing, and (**c**) peaking current source

input range, because a larger bias current at low V_{DC} will increase the comparator speed, and also generate a larger comparator offset for delay compensation.

To realize a bias current that is inversely proportional to V_{DC}, one may set the peak current point to be the lowest V_{DC} in the application, such as the I_{B1} in Fig. 4.16c. However, due to the square relation, the bias current will drop too much at the highest V_{DC}. Therefore, a dual-peaking current biasing circuit was proposed [3, 14]. The resistor R_1 in Fig. 4.16a is split into R_{1A} and R_{1B}. Two output currents (I_{B1} and I_{B2}) can be set with different peak current points and be summed together through M_{P1}. In doing so, no additional current branch is needed. The peak current point of I_{B1} (V_{PEAK1}) is set to be near the lowest V_{DC}, while the peak current point of I_{B2} (V_{PEAK2}) is set to be near the higher end of the V_{DC} range. I_{B2} is used to compensate I_{B1} at the high end of the V_{DC} range when it drops significantly. The peak currents of I_{B1} and I_{B2} can be calculated by

$$I_{B1,PEAK} = \frac{1}{2}k_n\left(\frac{W}{L}\right)_2\left[\sqrt{\frac{2I_{11}}{k_n(W/L)_1}} - I_{11}(R_{1A}+R_{1B})\right]^2, \tag{4.17}$$

$$I_{B2,PEAK} = \frac{1}{2}k_n\left(\frac{W}{L}\right)_3\left[\sqrt{\frac{2I_{12}}{k_n(W/L)_1}} - I_{12}R_{1A}\right]^2, \tag{4.18}$$

where I_{11} is the value of I_1 at V_{PEAK1}, and I_{12} is the value of I_1 at V_{PEAK2}.

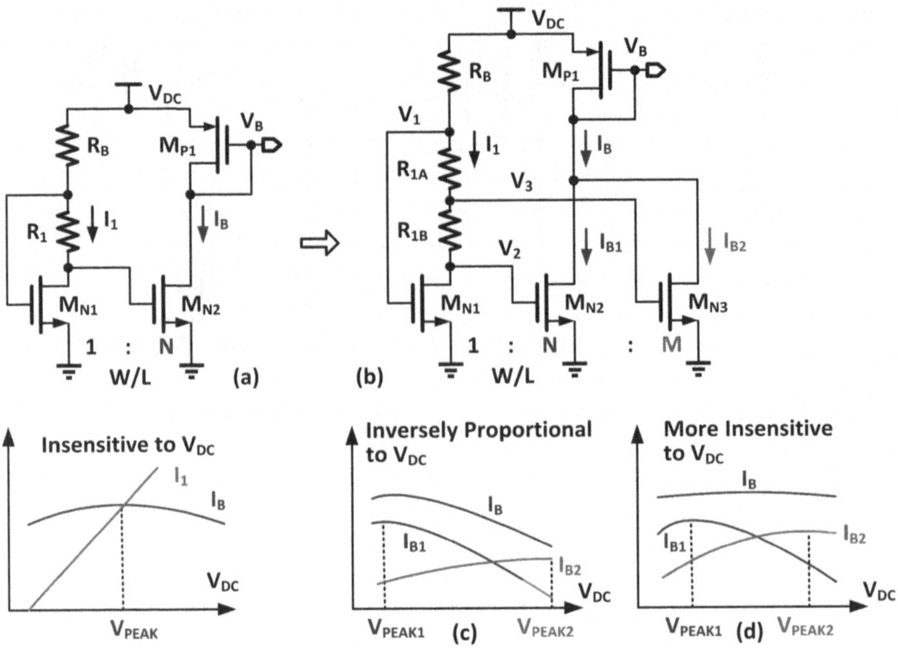

Fig. 4.16 Schematics of (**a**) the peaking current source, and (**b**) the dual-peaking current biasing circuits for (**c**) inversely proportional to supply bias current and (**d**) more insensitive to supply bias current

Basically, by tuning the peaking points of these two currents and the transistor sizes of M_{N2} and M_{N3}, the final output bias current I_B can be designed to many other shapes if needed. Another benefit of using the PCS is to have a well-controlled quiescent current over a wide supply range.

4.3.4 Full-Wave Rectifier Design Examples

Two full-wave active rectifiers with the Case 2 switched-offset comparator delay compensation scheme have been designed, implemented, and measured in a 0.35 μm CMOS N-well process. The first one (labeled as Rec1) [10] used a peaking current source, while the second one (Rec2) [14] used the proposed quasi-inversely proportional to V_{DC} (QIPV) current bias. Detailed analysis, optimization and measurement results will be presented in this sub-section.

The schematic of the NMOS active diode is shown in Fig. 4.17. Time-varying offset is introduced to the comparator by dynamic switched biasing currents from M_9 and M_{10} that are implemented in a push-pull fashion.

Note that V_{SW} of CMP1 can be equal to "1" only when V_{GN1} is "0" and V_{GN2} is "1". Assume that V_{GN2} is "1" in the previous phase such that V_{SW} is also "1", and

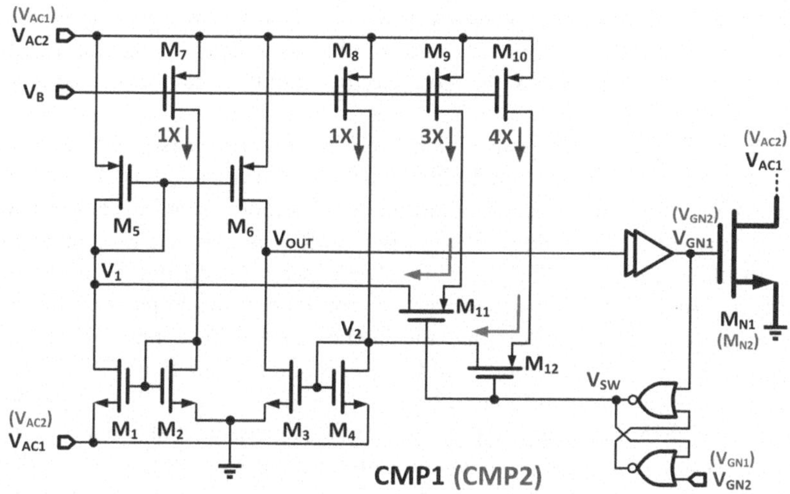

Fig. 4.17 The active diode 1 (2) with the Case 2 delay compensation scheme

the switches M_{11} and M_{12} are turned off. In the present phase, V_{GN1} drives V_{SW} low and turns on M_{11} and M_{12}, allowing auxiliary bias currents from M_9 and M_{10} to introduce the designed DC offset. This offset voltage of the differential pairs with unbalanced bias currents is

$$\begin{aligned} V_{OS} &= \sqrt{\frac{2I_+}{k_n(W/L)}} - \sqrt{\frac{2I_-}{k_n(W/L)}} \\ &= (\sqrt{n}-1)\sqrt{\frac{2I_-}{k_n(W/L)}} = (\sqrt{n}-1)V_{OV-}, \end{aligned} \tag{4.19}$$

where $k_n = \mu_n C_{ox}$, I_+ and I_- are the unbalanced bias currents, n is the ratio of I_+ and I_-, and V_{OV-} is the gate overdrive voltage of M_1 through M_4 when they are operating with balanced bias current (1X). The sizes of the common-gate input pairs M_1 through M_4 are the same, with V_{OV-} lower than 100 mV. The bodies of M_1 through M_4 are all connected to the on-chip ground, such that the threshold voltage V_{THN} of M_1 and M_4 would be smaller than V_{THN} of M_2 and M_3 when $V_{AC1} < 0$ due to the body effect. The bias currents of M_7 through M_{10} are 1X, 1X, 3X and 4X respectively. The current of M_{10} should be 3X (same as M_9); however, taking the body effect of M_4 into consideration, it is set as 4X instead to make V_2 even higher. The auxiliary bias currents also serve as slew-rate enhancement currents that charge up V_1 and V_2, starving M_6 and feeding M_1. Hence, V_{OUT} is pulled low, turning off the power switch M_{N1} right before $V_{AC1} > 0$ to prevent the occurrence of reverse current. The NOR SR-latch is added at each output of the comparators to avoid the aforementioned multiple-pulsing problem that can be caused by the dynamic offset scheme. Due to the NOR implementation of the SR-latch, V_{SW} stays low even when V_{GN1} goes low, keeping the artificial offset operative and preventing M_{N1} from turning on again in the same phase even if V_{AC1} is still lower than 0 V, and the multiple-pulsing problem is thus avoided. Another one-shot per cycle configuration used in [16] that pulls down the comparator output is not more reliable for eliminating multiple pulsing.

To save static power, the power supplies of the comparators CMP1 and CMP2 can be connected to (the distorted sinusoidal waveforms) V_{AC2} or V_{AC1}, respectively, such that only one comparator has the bias current to work in each half-cycle [8]. In addition, the V_{OUT} node of the comparators is coupled to ground in the layout, to make sure that this high-impedance node would not jump when the supply (V_{AC2} or V_{AC1}) is low. However, to eliminate multiple pulsing, the power supply of the gate-drive buffers and the latches should be connected to the DC output V_{DC}. All N-wells are connected to V_{DC}. Some PN junctions (P+ active area to N-well) are slightly forward biased by $V_{AC2} - V_{DC}$ (or $V_{AC1} - V_{DC}$, equal to V_{DS} of the power PMOS that is approximately 70 mV), the associated leakage current is negligible. When V_{AC2} or V_{AC1} is slightly higher than V_{DC} during the conduction time, the matched currents of M_7 through M_{10} would be larger than the designed bias current with larger V_{GS}. This arrangement is good for the comparators to have more instantaneous current and faster response.

The measured biasing current of the peaking current source used in Rec1 is shown in Fig. 4.18a, which matches well with simulation results. The peak current is designed to be around 4.1 μA when the supply voltage is 2.3 V, and the minimum bias current is 3.2 μA when $V_{DC} = 3.8$ V. When V_{DC} changes from 1.2 to 3.8 V, I_1 changes by over ±50%, but the bias current changes by only ±12.3%.

Simulated and measured results of the proposed QIPV bias circuit are shown in Fig. 4.18b. The size of M_{N3} is the same as M_{N1}, while $(W/L)_2$ is still four times of $(W/L)_1$, because I_1 at V_{PEAK2} is roughly two times higher than I_1 at V_{PEAK1}. With this assignment, $I_{B2,PEAK}$ is about half of $I_{B1,PEAK}$. The combined bias current $I_B = I_{B1} + I_{B2}$ then resembles a QIPV output current. The measured I_{B1} peaks at around 1.6 V with a current of 2.68 μA, and drops to 0.5 μA at $V_{DC} = 4$ V. The measured I_{B2} peaks at around 2.9 V with a current of 1.53 μA. Meanwhile, I_B peaks at around 1.7 V with 3.95 μA, and drops to 1.97 μA at $V_{DC} = 4$ V. The dual-peaking current biasing circuit can be easily modified to bias an amplifier designed to have a constant (or adaptive) bandwidth, for example, to compensate for the negative effects due to decreasing supply voltage.

4.3.4.1 Power Transistor Sizing

NMOS transistors are chosen to implement the active diodes and PMOS transistors to implement the cross-coupled pair for two reasons. First, by using PMOS transistors for the cross-coupled pair, they are driven by the AC input, not by the comparator. Thus, their parasitic gate capacitors do not affect the speed and the switching loss of the rectifier, as the gate capacitors are now part of the LC resonant tuning capacitor C_2 that recycles the charge, and the energy is just transferred between C_2 and the secondary coil L_2.

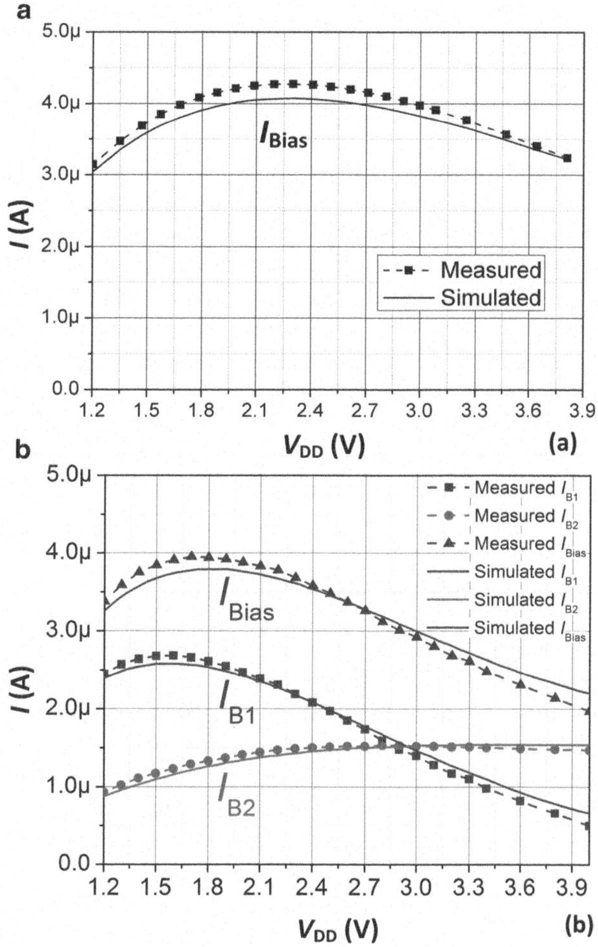

Fig. 4.18 Simulated and measured output current versus supply voltage of (**a**) the peaking current source for Rec1 and (**b**) the QIPV biasing for Rec2

Second, the mobility of NMOS transistors is higher, and results in smaller *W*/*L* ratios that reduce switching loss. For the trade-off between switching loss and conduction loss, W_N is set to be 600 μm with minimum channel length; otherwise, *W*/*L* of the power PMOS has to be large to achieve a small turn-on voltage drop ($W_P = 4000$ μm). In this design, the turn-on voltage drops of power NMOS V_{DSN} and PMOS $|V_{DSP}|$ are set to be roughly 250 mV and 70 mV, respectively, at $|V_{AC}| = 1.5$ V and $R_L = 500\ \Omega$. The total voltage drop is only about 320 mV in the worst case, which is much smaller than using passive diodes.

4.3.4.2 Start-Up Process

The lowest input amplitude $V_{AC,MIN}$ for our proposed rectifiers to work is determined by the minimum supply voltage of the comparators. The rectified DC voltage should be higher than $V_{GS} + V_{DSAT}$ for the comparators to work, so $V_{AC,MIN}$ is given by

$$V_{AC,MIN} = V_{DC,MIN} + V_{Drop} = V_{GS} + V_{DSAT} + V_{DSN} + |V_{DSP}|, \qquad (4.20)$$

where V_{DSN} and $|V_{DSP}|$ are the drain-to-source voltages of the power transistors. However, the lowest input amplitude for start-up is the same as that of using passive diodes, because when the active rectifier is relaxed, the DC output voltage V_{DC} is 0 and is unable to turn on the power NMOS switches. Instead, as illustrated in Fig. 4.19, parasitic diodes of the power NMOS switches will be forward-biased when $V_{AC} > V_{DC}$ during start-up. After the output capacitor is charged up to higher than the minimum supply voltage required by the comparators that are biased by the QIPV current source, the power NMOS switches are then activated, and the active rectifier could then function as designed. Latch-up problem could be avoided by careful layout. In more complicated control schemes, a start-up circuit may be needed to obtain a smooth transition or avoid malfunction.

4.3.4.3 Measurement Results

The proposed active rectifiers were designed and fabricated in a 0.35 μm CMOS N-well process. Micrographs of the rectifiers are shown in Fig. 4.20. The sizes of these two rectifiers, including the pads, are 0.12 mm^2 and 0.19 mm^2, respectively; and the active areas are 0.041 mm^2 and 0.065 mm^2, respectively. The measurement setup is shown in Fig. 4.21, including the planar coupling coils that are etched on single-side printed circuit boards (PCBs). The primary and secondary coils each have three turns with inner and outer radii of 0.75 cm and 1 cm respectively, and were separated by 1 cm during measurements. The measured inductance of the coils is 310 nH, and the series resistance is 480 mΩ at 13.56 MHz and 190 mΩ at DC. The DC output of the rectifier drives a 1.5 nF off-chip filtering capacitor.

Figure 4.22 shows the measured AC input and DC output waveforms of the proposed active rectifiers Rec1 and Rec2, respectively. The peak of I_{B1} is eventually designed to be located at 1.6 V instead of 1.2 V to reserve adequate margin for keeping the PCE high in the middle and higher range of V_{AC}. The optimized case is when the input amplitude V_{AC} is 3 V and R_L is 500 Ω. The input voltage ripples across V_{AC1} and V_{AC2} are due to the large input current changes during turning on and off the power NMOS switches. As can be observed from Fig. 4.22a and 4.22b, reverse current is well eliminated, which means that the NMOS switch is turned off when its drain voltage is higher than the ground voltage. The worst operating condition for the rectifier is at heavy load and low input amplitude, as shown in Fig. 4.22c and 4.22d. In this case, a small amount of reverse current is observed due to slower response time of the rectifier at lower supply voltages. Conduction loss of the power transistors increases at heavy load and when the gate overdrive voltage is small at low V_{AC}. The performance could further be improved if the bias current of the comparator is slightly increased to have a larger offset for faster response.

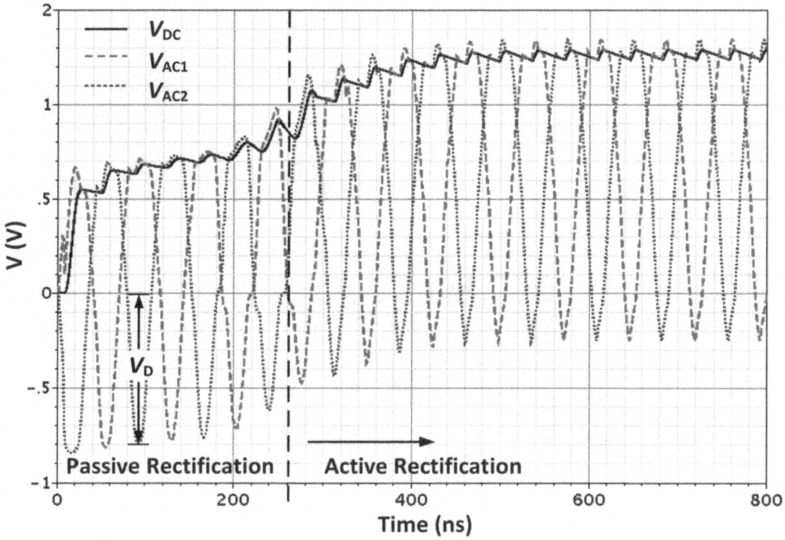

Fig. 4.19 Simulated start-up process of the active rectifier (Rec2)

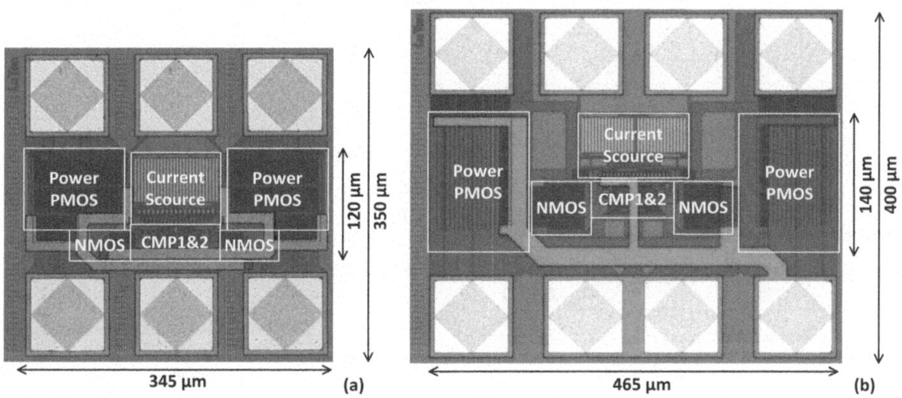

Fig. 4.20 Micrographs of (**a**) Rec1 and (**b**) Rec2

Figure 4.23 summarizes the measured voltage conversion ratios M of the rectifiers versus V_{AC} under different loading conditions (R_L = 500 Ω and 5 kΩ, respectively). The peak voltage conversion ratios of Rec1 and Rec2 are 92.3% and 93.1%, respectively, when R_L = 5 kΩ. The minimum M of Rec1 and Rec2 are 74% and 79%, respectively, when R_L = 500 Ω.

For PCE measurement, as shown in Fig. 4.24, a 10 Ω resistor is inserted in the input path to measure the AC input current I_{AC}. C_{2B} is used to filter the distorted V_{AC} waveforms caused by the 10 Ω resistor during large dI/dt transients. The data of V_{AC} and I_{AC} can be collected by two identical differential probes with the setup shown in Fig. 4.24a; or by two identical single-ended probes as shown in Fig. 4.24b.

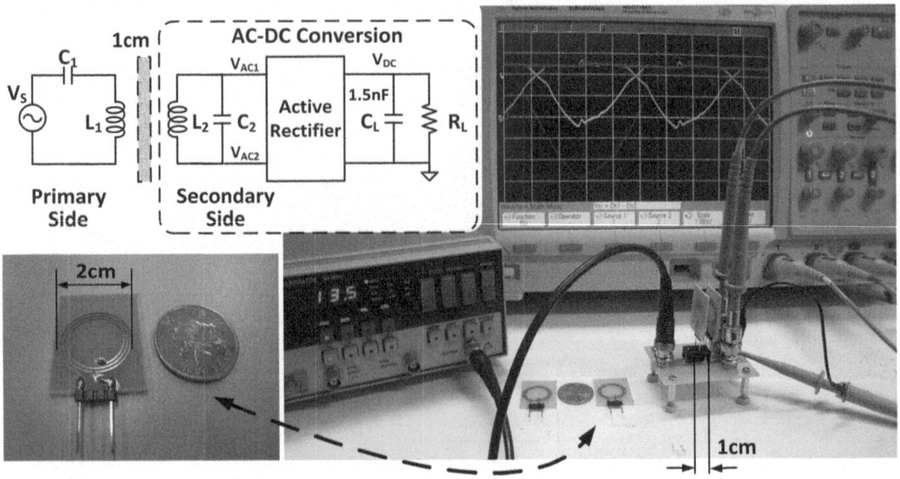

Fig. 4.21 Measurement setup for the 13.56 MHz active rectifiers

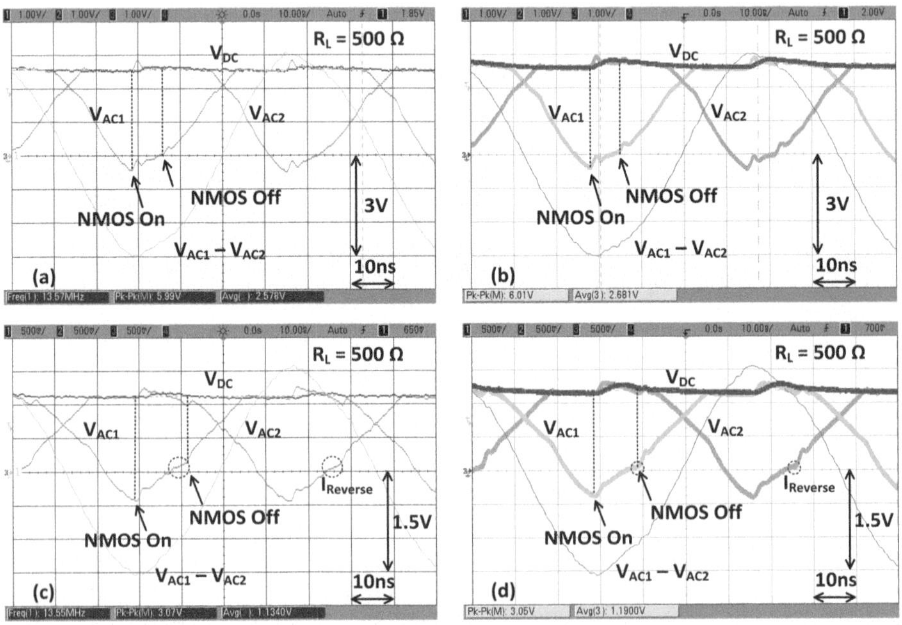

Fig. 4.22 Measured waveforms of AC inputs and DC output at $R_L = 500\Omega$ and $V_{AC} = 3$ V of (a) Rec1 and (b) Rec2; at $R_L = 500\ \Omega$ and $V_{AC} = 1.5$ V of (c) Rec1 and (d) Rec2

In addition, a voltage meter with floating terminals not connected to the ground (a handheld digital multimeter, for example) is needed for the measurement with single-ended probes. Note that according to (4.5), the PCE is not necessarily lower than the voltage conversion ratio at the same loading condition, as the highest PCE

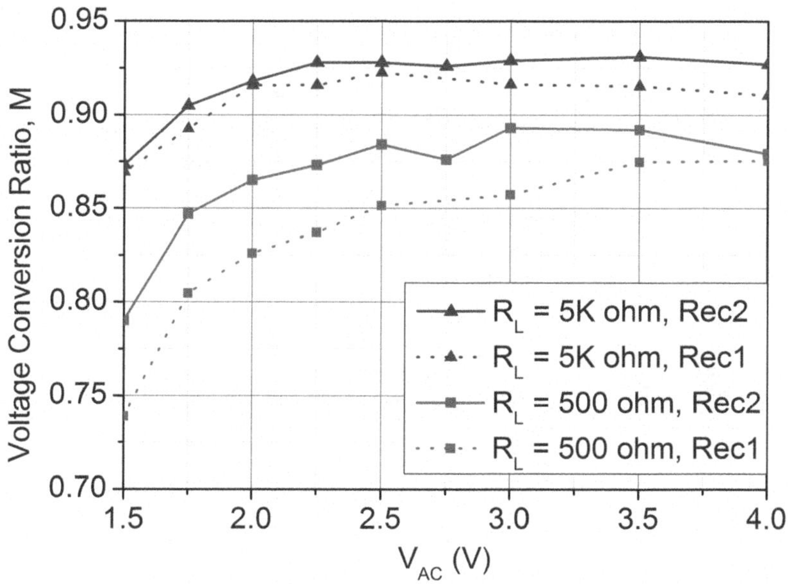

Fig. 4.23 Measured voltage conversion ratios of Rec1 and Rec2 with different loadings

usually is designed to be at heavy load, while the highest M is always obtained at light load.

In our experiments, two differential active probes with 1 GHz bandwidth were used to measure V_{AC} and I_{AC}. About 10,000 points of V_{AC} and I_{AC} (6 cycles) were integrated and averaged for P_{IN} in each PCE calculation. The secondary coil L_2 resonated with $C_{2A} + C_{2B}$ at 13.56 MHz, with $C_{2B}/C_{2A} = 0.15$. As shown in Fig. 4.25, the measured I_{AC} is the sum of I_{Cap} that passes through C_{2B} and I_{Rec} that flows into the rectifier. To obtain accurate PCE results, C_{2B}/C_{2A} cannot be large, otherwise, the large I_{CAP} of C_{2B} that does not dissipate power will affect the accuracy of the relatively small rectifier input current I_{REC} that dissipates power. Note also that the scales of the two identical probes should be kept the same to guarantee accuracy.

As summarized in Fig. 4.26a, with $R_L = 500\ \Omega$, the PCEs of Rec2 were measured to be 82.2–90.1% with $|V_{AC}|$ that varied from 1.5 to 4 V; while in Fig. 4.26b the PCEs of 82.3% and 71.2% were measured at $|V_{AC}| = 3$ V with $R_L = 100\ \Omega$ and 5 kΩ, respectively. Simulated PCEs are also included for reference: 84.2–90.7% were obtained with a load resistor of 500 Ω, and 82.9–81.3% were obtained at $|V_{AC}| = 3$ V with $R_L = 100\ \Omega$ and 5 kΩ, respectively. All the above results were measured and simulated at an input frequency of 13.56 MHz. In Fig. 4.26c, measured and simulated PCEs of Rec2 that operated at the transmission frequencies from 10 to 20 MHz are shown, with $R_L = 500\ \Omega$ and $|V_{AC}| = 3$ V. The measured PCEs matched quite well with the simulated PCEs at heavy load, and started to deviate from the simulated PCEs at light load of 1 kΩ since I_{REC} became too low to be measured accurately.

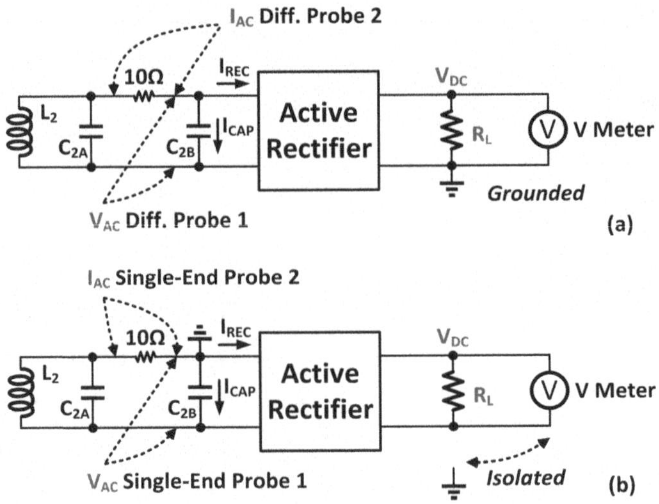

Fig. 4.24 The PCB schematics for PCE measurements with (**a**) differential voltage probes or (**b**) single-ended probes

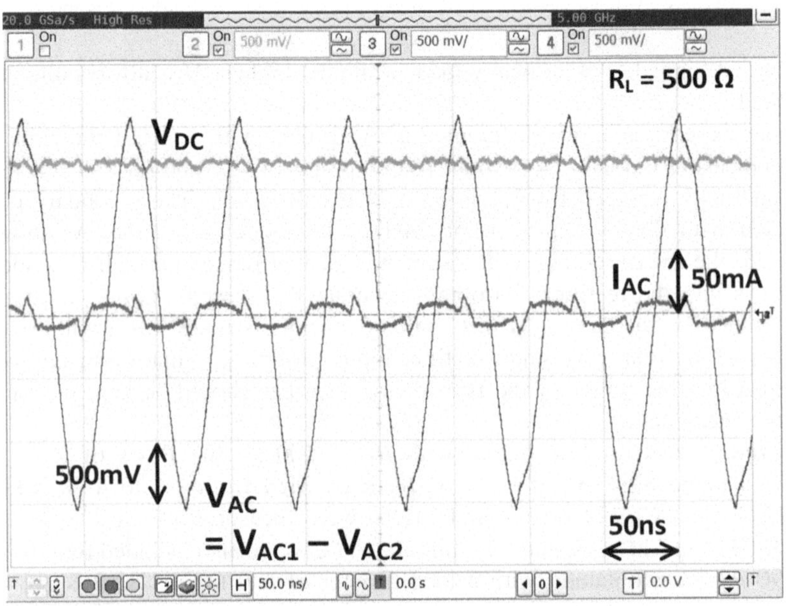

Fig. 4.25 Measured waveforms of V_{AC}, I_{AC} and V_{DC} for PCE calculation

Table 4.1 summarizes and compares the performance of our works with state-of-the-art designs. With inductively-coupled air coils operating at 13.56 MHz in the ISM frequency band, the proposed rectifier achieved good voltage conversion ratio and power conversion efficiency over a wide input range and loads.

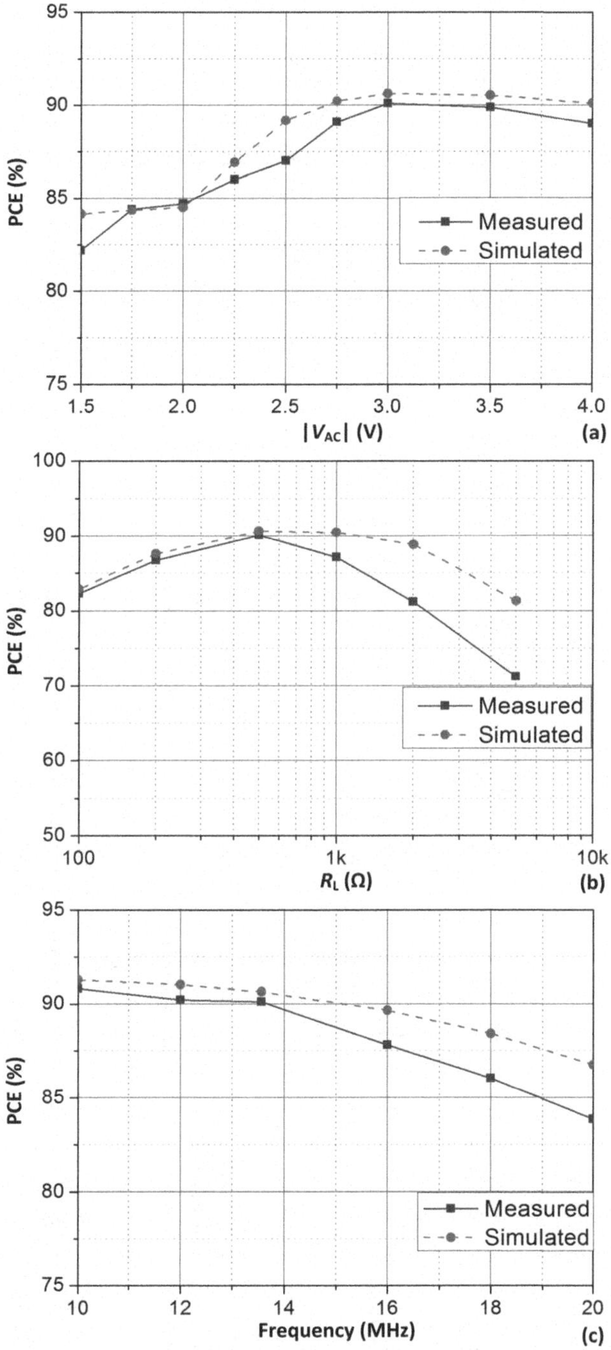

Fig. 4.26 Measured and simulated PCEs of the proposed Rec2 operating at 13.56 MHz with (**a**) $R_L = 500\ \Omega$ and (**b**) $|V_{AC}| = 3$ V; and (**c**) its frequency response with condition of $R_L = 500\ \Omega$ and $|V_{AC}| = 3$ V

Table 4.1 Comparison to the state-of-the-art rectifiers

	TCAS-II 06 [8]	JSSC 09 [9]	TCAS-I 11 [11]	TBCAS 12 [36]	JSSC 12 [13]	This work Rec1 [10]	This work Rec2 [14]
Technology	0.35 μm	0.35 μm	0.5 μm	0.18 μm	0.18 μm	**0.35 μm**	**0.35 μm**
Chip area	0.107 mm^2	1.03 mm^2	0.263 mm^2	0.608 mm^2	0.34 mm^2	**0.12 mm^2**	**0.19 mm^2**
Frequency	13.56 MHz	1.5 MHz	13.56 MHz	10 MHz	13.56 MHz	**13.56 MHz**	**13.56 MHz**
Input Amp.	1.5 ~ 3.5 V	1.2 ~ 2.4 V	3.3 ~ 5 V	0.8 ~ 2.7 V	N/A	**1.5 ~ 4 V**	**1.5 ~ 4 V**
Load Cap.	200 pF	1 μF	10 μF	200 pF	5.8 μF	**1.5 nF**	**1.5 nF**
Output voltage	1.2 ~ 3.22 V (R_L = 1.8 kΩ)	1.13 ~ 2.28 V (R_L = 2 kΩ)	2.5 ~ 3.9 V (R_L = 500 Ω)	0.3 ~ 2.0 V (R_L = 2 kΩ)	1.2 V (w/load chip)	**1.28 ~ 3.65 V (R_L = 1.8 kΩ)**	**1.28 ~ 3.56 V (R_L = 1.8 kΩ)**
		0.98 ~ 2.08 V (R_L = 100 Ω)				**1.13 ~ 3.50 V (R_L = 500 Ω)**	**1.19 ~ 3.52 V (R_L = 500 Ω)**
Coil Diam.	6.0 cm	N/A	3.0 cm	N/A	7 mm	**2.0 cm**	**2.0 cm**
P_{OUT} (Max.)	5.76 mW	43.3 mW	30.42 mW	2 mW	112.5 mW	**24.5 mW**	**24.8 mW**
Voltage conversion ratio, M	0.78 ~ 0.92 (R_L = 1.8 kΩ)	0.94 ~ 0.95 (R_L = 2 kΩ)	0.76 ~ 0.81 (R_L = 500 Ω)	0.6 ~ 0.89 (R_L = 2 kΩ)	N/A	**0.85 ~ 0.9 (R_L = 1.8 kΩ)**	**0.873 ~ 0.93 (R_L = 1.8 kΩ)**
		0.82 ~ 0.84 (R_L = 100 Ω)				**0.74 ~ 0.88 (R_L = 500 Ω)**	**0.79 ~ 0.89 (R_L = 500 Ω)**
PCE (simulated)	65% ~ 89% (R_L = 1.8 kΩ)	82% ~ 87% (R_L = 100 Ω)	71% ~ 84.5% (R_L = 500 Ω)	60% ~ 86% (R_L = 2 kΩ)	93% (w/load chip)	**82.5% ~ 89.6% (R_L = 500 Ω)**	**84% ~ 90.7% (R_L = 500 Ω)**
PCE (measured)	N/A	N/A	68% ~ 80.2% (R_L = 500 Ω)	37% ~ 80% (R_L = 2 kΩ)	N/A	**N/A**	**82% ~ 90.1% (R_L = 500 Ω)**

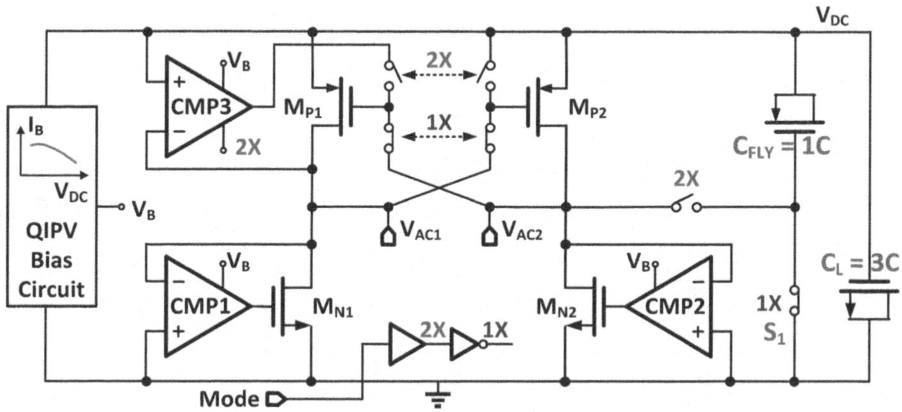

Fig. 4.27 Fully integrated 1X/2X active rectifier with QIPV bias current

4.3.5 Reconfigurable Rectifier Design Example

To cater for coupling coefficient k variation with distance and/or orientation, as discussed in Chap. 2, a reconfigurable rectifier may be used to increase V_{DC} without increasing the transmitted power under certain conditions. A fully integrated reconfigurable active rectifier (labeled as Rec3) [3] that can be switched between the full-wave rectifier (1X) mode and the voltage doubler (2X) mode in the 30 mW range with all capacitors fabricated on-chip is shown in Fig. 4.27.

In the traditional reconfigurable rectifier shown in Fig. 4.2, the output capacitors are connected in series which is not area-efficient for on-chip implementation. The effectiveness of utilizing on-chip capacitors in this design is significantly improved by the switching arrangement that avoids connecting the output capacitors in series in the 2X mode. Reverse current is reduced for $|V_{AC}|$ that ranges from 1.25 V to 4 V by the bias current that is quasi-inversely proportional to the output DC voltage V_{DC} (QIPV).

Figure 4.28 shows the conventional and the proposed arrangement of C_L. Assume that the total capacitance available is $4C$ for implementing C_L, in the conventional topology, two $2C$ capacitors are connected in series that results in $C_{L,EQ}$ equal to C in both the 1X and the 2X mode. In the proposed topology, $C_{FLY} = C$ and $C_L = 3C$ are connected through switches S_1 and S_2. In the 1X mode, S_1 is turned on, the two capacitors are in parallel, and $C_{L,EQ}$ is $4C$; and in the 2X mode, S_1 is opened, C_{FLY} acts as a flying capacitor to charge up C_L, and $C_{L,EQ}$ is $3C$. As a result, the proposed topology increases $C_{L,EQ}$ by 4 and 3 times compared to the universal rectifier in the 1X and the 2X mode, making it possible to integrate the capacitors on-chip. In our case, C_{FLY} and C_L are equal to 1 and 3 nF respectively. Note that C_{FLY} serves as a pumping capacitor for voltage doubling in a charge pump, and a small C_{FLY} will limit the VCR and consequently the PCE.

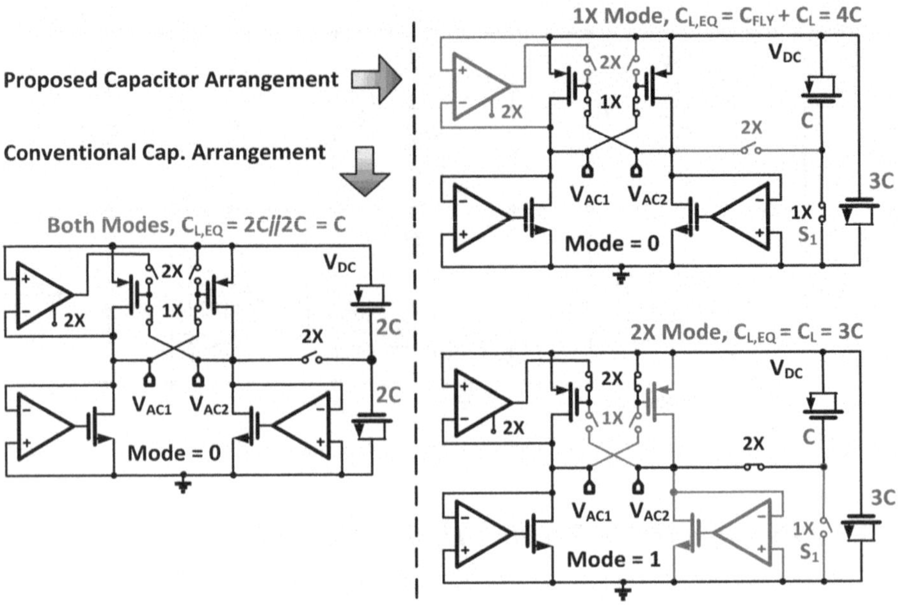

Fig. 4.28 The conventional and the proposed arrangement of C_L, assuming that the total capacitance available is $4C$ for implementing C_L

Based on the considerations of the VCR at the maximum loading condition in the 2X mode, C_{FLY} is set to be 1 nF.

The schematics of the comparators CMP1 (CMP2) with M_{N1} (M_{N2}), and the comparator CMP3 with M_{P1}, are shown in Fig. 4.29. The "Mode" signal also changes the value of the unbalanced currents in CMP1 and CMP2, such that the offsets of CMP1 and CMP2 are set to different values in the two modes, because the input sinusoidal wave has different slew rates in the 1X and the 2X mode. Note that for Rec3 the minimum value of V_{DC} is 1.6 V as the rectifier would operate in the 2X mode in low-voltage conditions. Thus, the proposed QIPV I_{BIAS} helps the rectifier to reduce the reverse current for V_{DC} from 1.6 to 4 V.

The comparator CMP3 is disabled in the 1X mode by cutting off its bias current with an "En" signal, because the power PMOS transistors are cross-coupled in the 1X mode. In the 2X mode, CMP2 is not disabled even it does not need to work in this mode, because its input terminal V_{AC2} is always equal to roughly half of V_{DC} (always above Ground voltage) in the 2X mode. It means that the CMP2 output would not fluctuate in the 2X mode, and no extra switching loss will be induced by CMP2 and M_{N2}.

In designing the transistor sizes of the power MOS, we need to consider the tradeoff for peak VCRs and PCEs between the 1X mode and the 2X mode. Ideally with a resistive load, the output voltage V_{DC} in the 2X mode is double of the $|V_{AC}|$ input, and therefore the input current of the converter is double of the output current. Thus, for the same load condition, larger current needs to be conducted

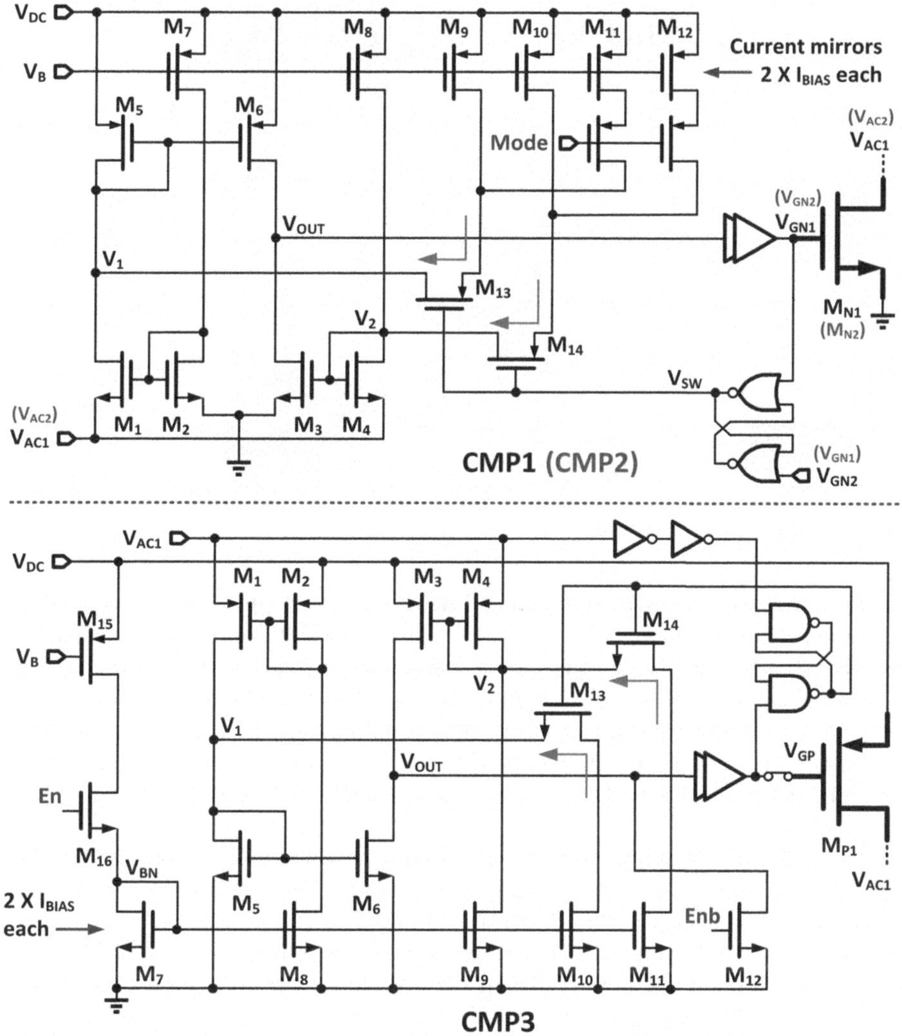

Fig. 4.29 Schematics of the CMP1 (CMP2) with M_{N1} (M_{N2}), and the CMP3 with M_{P1}

in the 2X mode, and consequently larger transistors are needed for the 2X mode. Another point is that, as mentioned in Sect. 4.3.4.1, when the PMOS transistors are configured as cross-coupled pair, they are driven by the AC input, not the comparator, and their parasitic gate capacitors do not affect the speed and the switching loss of the rectifier, as they are part of the LC resonant tuning capacitor C_2 that do not dissipate power. However, for the 1X/2X reconfigurable rectifier, the power PMOS M_{P1} is driven by CMP3 through a buffer in the 2X mode, and cannot be too large as for the full-rectifiers Rec1 and Rec2.

Fig. 4.30 Chip micrograph of Rec3

The proposed 1X/2X rectifier was fabricated in a 0.35 μm CMOS process. The chip micrograph is shown in Fig. 4.30. The active area is 0.1 mm^2, and the capacitor area is 1.3 mm^2. The flying and output capacitors C_{FLY} and C_L are MOS capacitors with a capacitance density of 3.2 fF/μm^2 that could be much higher for an advanced process with stacked metal capacitors. The coupling coils L_1 and L_2 for measurements are 2 cm and 1.8 cm in diameter, respectively. The secondary inductor L_2 is 268 nH and resonates with the tuning capacitor $C_2 = C_{2A} + C_{2B}$ of 514 pF at 13.56 MHz.

The measured AC input and DC output voltage waveforms in both modes with $R_L = 500\ \Omega$ and $C_L = 4$ nF (on-chip) are shown in Fig. 4.31. The worst voltage conversion ratio occurs at the lowest $|V_{AC}|$ points.

VCRs and PCEs under different conditions are plotted in Fig. 4.32. With $R_L = 500\ \Omega$, the VCR is 0.85–0.9 in the 1X mode with the QIPV bias current optimized for this case, and is 1.3–1.61 in the 2X mode. With $R_L = 5\ k\Omega$, the VCR is 0.92–0.95 in the 1X mode, and is 1.73–1.77 in the 2X mode. With $R_L = 500\ \Omega$, the PCEs of the 1X and the 2X mode are measured to be 81–84.2% and 61–76%, respectively. Table 4.2 summarizes the performance of Rec1, Rec2, and Rec3 with the state-of-the-art full-wave and reconfigurable designs.

4.4 DLL-Based Rectifiers

Besides the mentioned analog solutions, there are other variations of CMOS active rectifiers that have been proposed [22–26] for near-field wireless power transfer. Figure 4.33 shows a WPT active rectifier that operates at a relatively high frequency

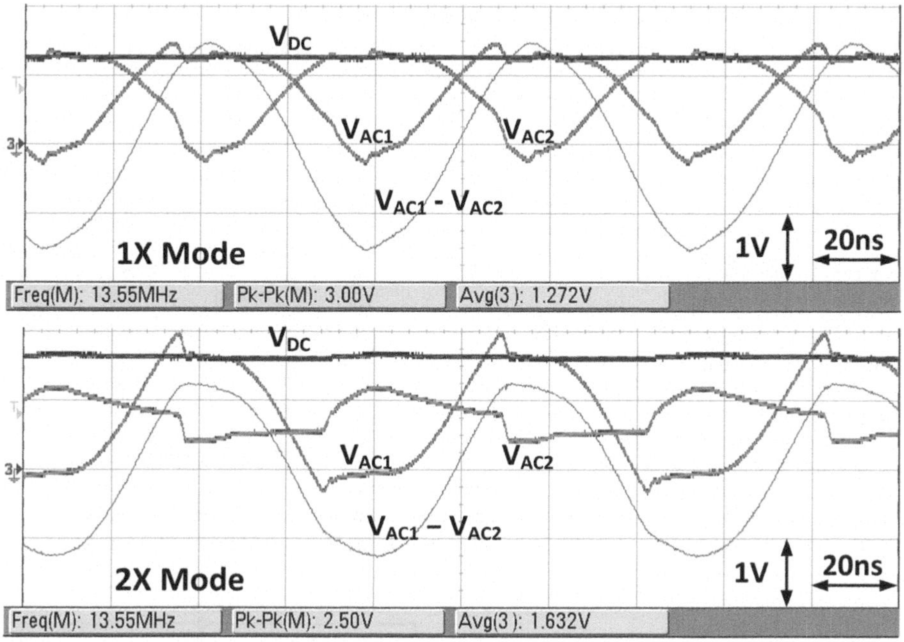

Fig. 4.31 Measured AC input and DC output voltage waveforms in both modes with $R_L = 500\ \Omega$ and $C_L = 4$ nF (on-chip) at their lowest $|V_{AC}|$ points

Fig. 4.32 Measured voltage conversion ratios and power conversion efficiencies of the 1X/2X rectifier in both modes

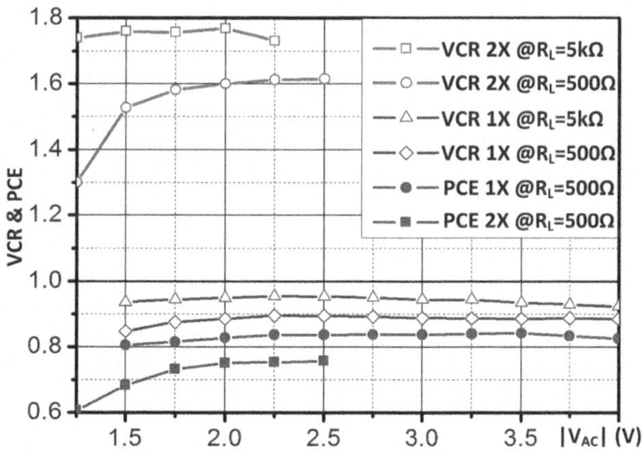

(150 MHz). It is a delay-locked-loop (DLL) based synchronous active rectifier that uses the DLL to synchronize the gate control signals with the AC input waveforms and to generate discrete timing slots for controlling the power switches. The clock signal "Clk" is recovered by comparing the AC input signals, and then it is used to generate multiple time-shifted pulses. The multiplexers (MUXs) are controlled by the comparators to relay the "set" and "reset" timing signals to the gate drive buffers.

Table 4.2 Comparison of full-wave and reconfigurable rectifiers

	[8]	[12]	[2]	Rec1 [10]	Rec2 [14]	Rec3 [3]
Tech. (μm)	0.35	0.18	0.5	**0.35**		
Area (mm^2)	0.107	0.009	0.585	**0.04**	**0.065**	**0.1(w/o C_L)**
Freq. (MHz)	**13.56**					
$\lvert V_{AC}\rvert$ (V)	1.5–3.5	0.9–2	3.2–5 (1X) 1.7–2.5 (2X)	**1.5–4**	**1.5–4**	**1.5–4 (1X)** **1.25–2.5 (2X)**
C_L (nF)	0.2	10,000	2000	**1.5**	**1.5**	**4**
R_L (Ω)	1.8 k	1 k	500	**500**	**500**	**500**
V_{DC} (V)	1.2–3.2	0.45–1.8	2.5–4.3	**1.13–3.5**	**1.2–3.5**	**1.27–4**
Coil (cm)	6.0	N/A	3.0	**2.0**	**2.0**	**1.8**
$P_{OUT,MAX}$ (mW)	5.76	3.2	37	**24.5**	**24.8**	**32**
VCR 1X	0.78–0.92	0.82–0.89	~0.84	**0.74–0.88**	**0.79–0.89**	**0.85–0.9**
VCR 2X	N/A	N/A	~1.41	**N/A**	**N/A**	**1.3–1.61**
PCE 1X	65–89%	60–82%	73–77%	**82–90%**	**82–90%**	**81–84%**
PCE 2X	N/A	N/A	64%–70%	**N/A**	**N/A**	**61–76%**

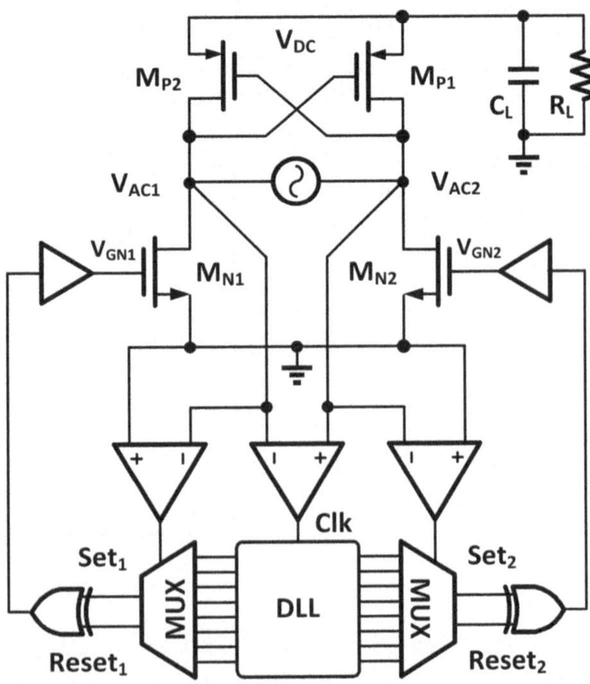

Fig. 4.33 A DLL-based synchronous active rectifier

In the open-loop finite resolution design of [22], the RF signal envelope is assumed to vary slowly and that small timing errors cause insignificant reduction of the rectifier efficiency. This assumption applies better to the series-resonant receiver than to the parallel-resonant receiver, as demonstrated in [16]. The reason is that, for the parallel-resonant receiver, the delay of the active diode increases the conduction time of the power transistors, resulting in extra conduction loss. Moreover, as the duration of the turn-off delay is comparable to the forward-current conduction time when operating at high frequencies, this extra conduction loss is significant and the efficiency degradation is severe. On the other hand, the series-resonant tank works as a current source, and the forward conduction time is much longer than that of the parallel-resonant receiver. Therefore, with the same amount of turn-off delay, the reverse current in the series-resonant receiver is only a small portion of the forward current, so the efficiency degradation is limited.

In another two LC series-resonant WPT designs [23, 24], replica delays were inserted in the DLL to imitate the gate-drive delay. By doing so, the reverse current can be well controlled when the replica delay matches well with the actual gate-drive delay. Alternatively, a sampled voltage-to-time conversion block was used in [25] to enable the power NMOS with zero-voltage switching (ZVS) and to eliminate the reverse current.

4.5 Rectifiers for RF Energy Harvesting

Although most of the WPT applications employ near-field operation for high and medium power levels, WPT with far-field operation has the advantage of longer transmission distance. In addition, RF energy harvesting together with other harvested ambient energy sources could serve as the power source for autonomous ultra-low-power internet-of-things (IoT) devices.

The majority of communication systems are designed to operate in the ultra-high-frequency (UHF) ISM bands (300 MHz–3 GHz), and the density of wireless devices keeps increasing rapidly worldwide in the recent decade. Thus, there is sufficient RF energy readily available in the environment in the UHF band. Moreover, using UHF for wireless power transfer could lead to a low-cost and/or small-size solution.

RF energy harvesting requires the rectifier to operate in the UHF range, and the rectification is commonly known as RF-DC conversion. While AC-DC conversion is usually discussed in the voltage domain, RF-DC conversion is mostly considered in the power domain. Since the RF input power level is usually low, multiple step-up stages are needed to attain a reasonably high output voltage. Figure 4.34 shows the voltage multiplier that is similar to the Dickson charge pump [27] used in [28] as a multi-stage rectifier for the RF to DC conversion at UHF. Each stage is a voltage doubler, and the ideal no-load output voltage of an N-stage converter is $2N$ times of the RF input amplitude.

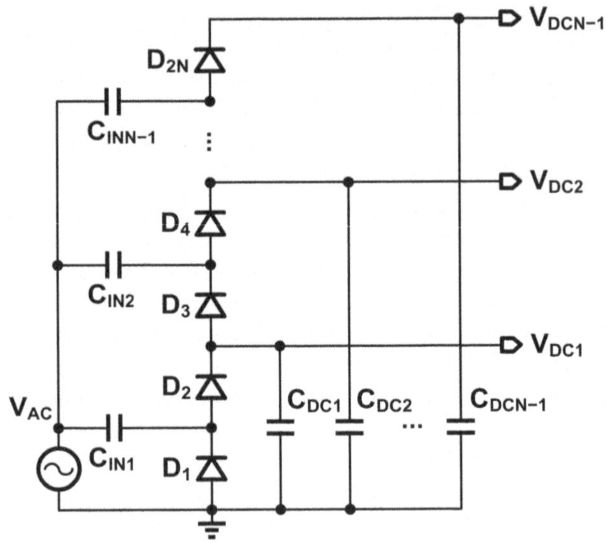

Fig. 4.34 Schematic of the Dickson-type multi-stage step-up rectifier

An interesting and useful characteristic of the multi-stage rectifier is that the number of stages N does not affect the maximum achievable PCE [29]. Note that each stage draws current from the RF input source in the charging phase of each cycle, and the stages are stacked in the discharging phase to achieve a high output voltage. In other words, the operation of each stage resembles that of a single-stage rectifier, and thus the maximum achievable PCE is basically independent of the number of stacking stages.

The appreciable diode drop is the main constraint that limits the low-voltage operation of the rectifiers, and this problem is exacerabated for RF wireless power transfer for longer distance transmission with very low coupling coefficient. Therefore, the cross-connected (CC) CMOS rectifier [30], as shown in Fig. 4.35, is commonly used for its low-voltage and auto-switching characteristics. Provided that the RF input voltage is high enough, the MOS transistors will act as switches with low on-resistance for rectification, and the forward voltage drop is minimized. However, a high leakage current may occur if the RF input voltage is too high because the PMOS and NMOS will be turned on simultaneously during the transition instants, similar to the shoot-through current problem of a CMOS inverter. To achieve better sensitivity and to achieve high heavy-load efficiency, the rectifier needs larger transistors, but will then cause more reverse leakage current that limits the efficiency at high input power. Larger transistors also increase the parasitic loss during the step-up conversion [31]. It has been demonstrated in [6] that the voltage conversion ratio (VCR) of the cross-connected topology could be larger than 0.8 when the input amplitude of the rectifier was higher than 150 mV.

To increase the received power, another solution that has much lower dynamic leakage current and low diode drop is to use a V_{TH} compensation voltage [33–36] to bias the diode-connected MOSFET, and the circuit is also shown in Fig. 4.35.

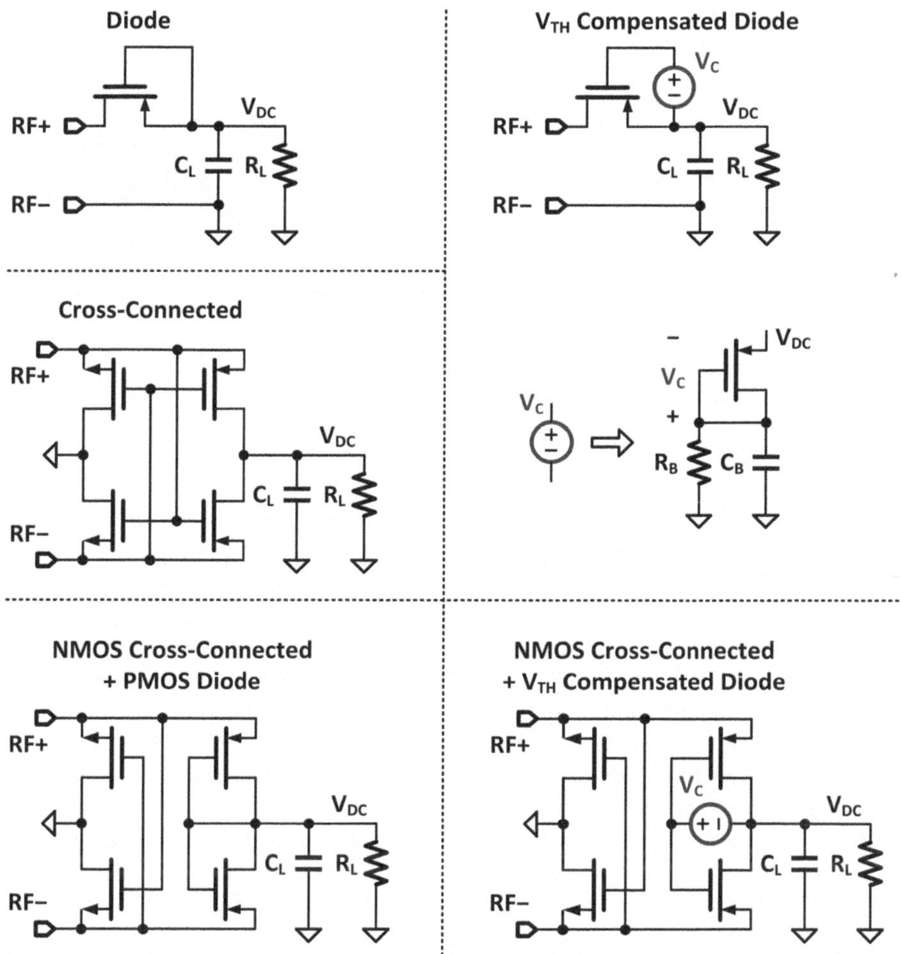

Fig. 4.35 Schematics of half-wave rectifiers, cross-connected rectifiers, and rectifiers with V_{TH} compensation

However, if the compensation voltage is realized by a tiny bias current (due to a restricted power budget) through a large resistor R_B filtered by a large capacitor C_B, then considerably large area would be consumed for a multi-stage rectifier.

Figure 4.36 shows the simulated PCEs of the three mentioned RF-DC rectifier topologies with $R_L = 100$ kΩ in a low-leakage 65 nm process. All three designs consist of three cascade stages for fair comparison. Obviously, the CC rectifier achieves the highest peak PCE among the three rectifiers. However, its high-PCE range is narrow due to the dynamic leakage current. A high-PCE range extension technique that consists of one high-power path and one low-power path with auto-selection can be used [32]. For the rectifier with cross-connected NMOS and V_{TH}-compensated PMOS, the performance is better than the diode-connected MOS

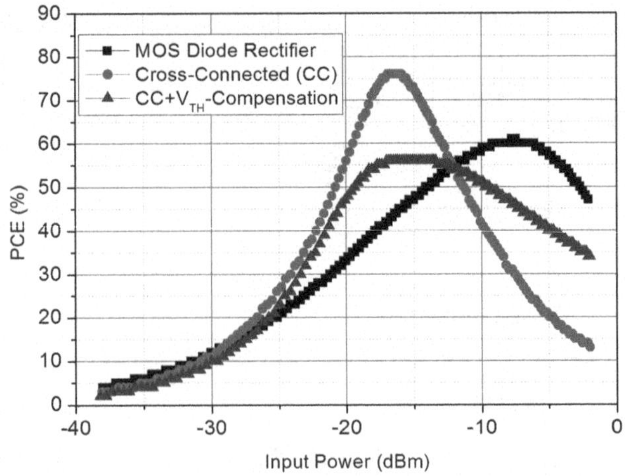

Fig. 4.36 Simulated PCEs of three rectifier topologies with $R_L = 100$ kΩ

rectifier at a lower input power range, and achieves a wider input power range than the CC rectifier. As shown in Fig. 4.37, the VCR for the CC topology will drop at high input-power levels, while that of the diode-based rectifiers are monotonic with respect to the input power.

An inter-stage V_{TH}-compensation scheme was introduced in [37, 38], and a 5-stage example is shown in Fig. 4.38. The V_{TH}-compensation was achieved by connecting the gate terminal to the output of a higher stage and as such, no large R_B and C_B have to be used while achieving similar performance. In fact, a similar strategy has been employed in step-up charge pump designs [39, 40].

Detailed analysis and simulation results show that the output voltage of a rectifier with the transistors operating in the subthreshold region is different from that of operating in the saturation region. As it has been analyzed in Sect. 4.2, the I-V characteristic of the transistors operating in the subthreshold region follows an exponential I_D/V_D relationship. With a typical value of ζ at room temperature, V_{GS} decreases by roughly 80 mV for I_D to decrease by one decade, which is independent of the value of V_{TH}. Therefore, it was shown that transistor size and V_{TH} have no effect on V_{DC}, and -32 dBm sensitivity was achieved with 50 stages [41].

4.6 Summary and Discussion

Passive and active rectifiers have been extensively discussed in this chapter. For passive rectifiers, using the composite CMOS diode results in low reverse leakage current and comparable forward current capability, which is favorable for low-power rectifiers.

The operation mechanism of the comparator-based active rectifiers for near-field WPT has been analyzed, and circuit design techniques and considerations have been discussed. In addition, a couple of design examples of active rectifiers with the

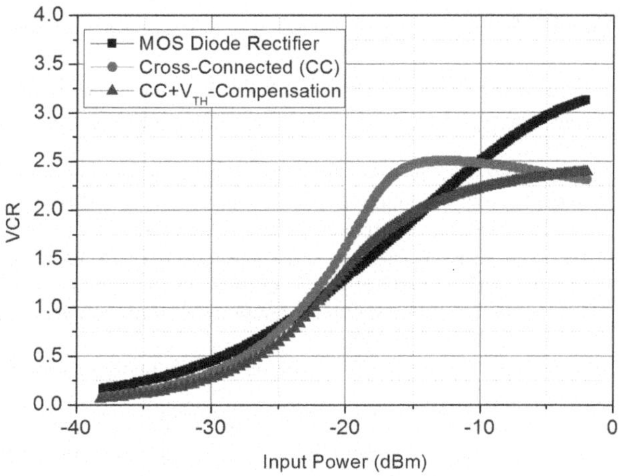

Fig. 4.37 Simulated VCRs of three rectifier topologies with $R_L = 100\ \text{k}\Omega$

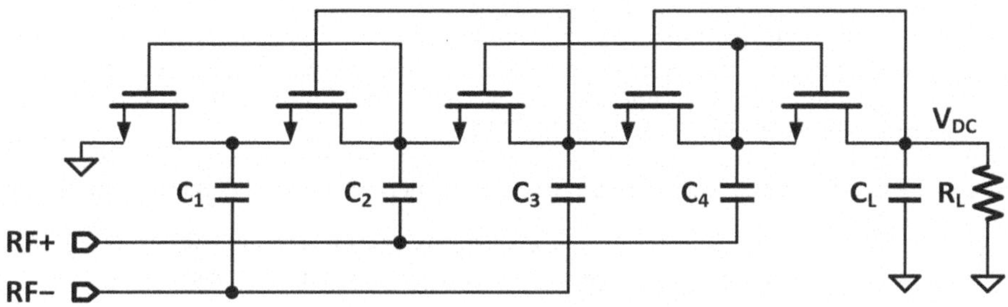

Fig. 4.38 Schematic of the inter-stage V_{TH}-compensation rectifier

switched-offset scheme are demonstrated. To optimally eliminate the reverse current associated with active diodes, the adaptive on- and off-delay compensation technique can be considered. Alternatively, DLL-based synchronous rectifiers have also been demonstrated to achieve promising performance, especially for the series-resonant receiver case.

For far-field WPT, voltage boosting is achieved by multi-stage rectifiers with structures that are similar to step-up charge pumps. The cross-connected topology was proven to be capable of achieving high efficiency at UHF, but it suffers from dynamic leakage current when the RF input amplitude is too high. The inter-stage V_{TH}-compensation scheme has also demonstrated promising performance, and more sophisticated inter-stage gate control of the switches might be a possible solution.

References

1. Ghovanloo M, Najafi K (2004) Fully integrated wideband high-current rectifiers for inductively powered devices. IEEE J Solid State Circuits 39:1976–1984. doi:10.1109/JSSC.2004.835822
2. Lee H-M, Ghovanloo M (2012) An adaptive reconfigurable active voltage doubler/rectifier for extended-range inductive power transmission. IEEE Trans Circuits Syst II Exp Briefs 59:481–485. doi:10.1109/TCSII.2012.2204840
3. Lu Y, Li X, Ki W-H, Tsui C-Y, Yue CP (2013) A 13.56MHz fully integrated 1X/2X active rectifier with compensated bias current for inductively powered devices. In: 2013 I.E. international solid-state circuits conference digest of technical papers (ISSCC), pp 66–67. doi:10.1109/ISSCC.2013.6487639
4. Razavi B (2000) Design of analog CMOS integrated circuits, 1st edn. McGraw-Hill, Inc., New York
5. Levacq D, Liber C, Dessard V, Flandre D (2004) Composite ULP diode fabrication, modelling and applications in multi-Vth FD SOI CMOS technology. Solid State Electron 48:1017–1025. doi:10.1016/j.sse.2003.12.016
6. Stoopman M, Keyrouz S, Visser HJ, Philips K, Serdijn W (2014) Co-design of a CMOS rectifier and small loop antenna for highly sensitive RF energy harvesters. IEEE J Solid State Circuits 49:622–634. doi:10.1109/JSSC.2014.2302793
7. Haddad PA, Gosset G, Raskin JP, Flandre D (2016) Automated design of a 13.56 MHz 19 μW passive rectifier with 72% efficiency under 10 μA load. IEEE J Solid State Circuits 51:1290–1301. doi:10.1109/JSSC.2016.2527714
8. Lam Y-H, Ki W-H, Tsui C (2006) Integrated low-loss CMOS active rectifier for wirelessly powered devices. IEEE Trans Circuits Syst II Exp Briefs 53:1378–1382. doi:10.1109/TCSII.2006.885400
9. Guo S, Lee H (2009) An efficiency-enhanced CMOS rectifier with unbalanced-biased comparators for transcutaneous-powered high-current implants. IEEE J Solid State Circuits 44:1796–1804. doi:10.1109/JSSC.2009.2020195
10. Lu Y, Ki W-H, Yi J (2011) A 13.56MHz CMOS rectifier with switched-offset for reversion current control. In: 2011 symposium on VLSI circuits (VLSIC), pp 246–247
11. Lee HM, Ghovanloo M (2011) An integrated power-efficient active rectifier with offset-controlled high speed comparators for inductively powered applications. IEEE Trans Circuits Syst I Regul Pap 58:1749–1760. doi:10.1109/TCSI.2010.2103172
12. Cha HK, Park WT, Je M (2012) A CMOS rectifier with a cross-coupled latched comparator for wireless power transfer in biomedical applications. IEEE Trans Circuits Syst II Exp Briefs 59:409–413. doi:10.1109/TCSII.2012.2198977
13. Chung H, Radecki A, Miura N et al (2012) A 0.025-0.45 W 60%-efficiency inductive-coupling power transceiver with 5-bit dual-frequency feedforward control for non-contact memory cards. IEEE J Solid State Circuits 47:2496–2504. doi:10.1109/JSSC.2012.2206686
14. Lu Y, Ki W-H (2014) A 13.56 MHz CMOS active rectifier with switched-offset and compensated biasing for biomedical wireless power transfer systems. IEEE Trans Biomed Circuits Syst 8:334–344. doi:10.1109/TBCAS.2013.2270177
15. C-Y W, Qian X-H, Cheng M-S et al (2014) A 13.56 MHz 40 mW CMOS high-efficiency inductive link power supply utilizing on-chip delay-compensated voltage doubler rectifier and multiple LDOs for implantable medical devices. IEEE J Solid State Circuits 49:2397–2407. doi:10.1109/JSSC.2014.2356459
16. Cheng L, Ki WH, Lu Y, Yim TS (2016) Adaptive on/off delay-compensated active rectifiers for wireless power transfer systems. IEEE J Solid State Circuits 51:712–723. doi:10.1109/JSSC.2016.2517119
17. Huang C, Kawajiri T, Ishikuro H (2015) A near-optimum 13.56 MHz active rectifier with circuit-delay real-time calibrations for high-current biomedical implants. In: 2015 IEEE custom integrated circuits conference, pp 1–4

18. Rabaey JM, Chandrakasan A, Nikolic B (2003) Digital integrated circuits: a design perspective, 2nd edn. Prentice-Hall, New Jersey
19. Frederiksen TM (1972) Constant current source. US Patent 3,659,121, 25 April 1972
20. Gray PR, Hurst P, Lewis S, Meyer RG (2009) Analysis and design of analog integrated circuits, 5th edn. Wiley, New York
21. Lo MY, Ki W-H, Mow WH (2009) A 20MHz switched-current input sample-and-hold circuit for current mode analog iterative decoders. In: IEEE International Symposium on Integrated Circuits, Singapore, pp 283–286, December 2009
22. Radecki A, Chung H, Yoshida Y, et al (2011) 6W/25 mm inductive power transfer for non-contact wafer-level testing. In: 2011 IEEE international solid-state circuits conference digest of technical papers (ISSCC), pp 230–232
23. Moon Y-J, Roh Y-S, Yoo C, Kim D-Z (2012) A 3.0-W wireless power receiver circuit with 75-% overall efficiency. In: 2012 IEEE Asian solid state circuits conference (A-SSCC), pp 97–100
24. Park H, Jang J, Kim H et al (2016) A design of a wireless power receiving unit with a high-efficiency 6.78-MHz active rectifier using shared DLLs for magnetic-resonant A4 WP applications. IEEE Trans Power Electron 31:4484–4498. doi:10.1109/TPEL.2015.2468596
25. Shinohara H, Miyaji K (2015) A ZVS CMOS active diode rectifier with voltage-time-conversion delay-locked loop for wireless power transmission. In: 2015 IEEE Asian solid-state circuits conference (A-SSCC), pp 1–4
26. Xu H, Lorenz M, Bihr U, et al (2014) Wide-band efficiency-enhanced CMOS rectifier. In: 2014 I.E. international symposium on circuits and systems (ISCAS), pp 614–617
27. Dickson JF (1976) On-chip high-voltage generation in MNOS integrated circuits using an improved voltage multiplier technique. IEEE J Solid State Circuits 11:374–378. doi:10.1109/JSSC.1976.1050739
28. Karthaus U, Fischer M (2003) Fully integrated passive UHF RFID transponder IC with 16.7-μW minimum RF input power. IEEE J Solid State Circuits 38:1602–1608. doi:10.1109/JSSC.2003.817249
29. Yi J, Ki W-H, Tsui C-Y (2007) Analysis and design strategy of UHF micro-power CMOS rectifiers for micro-sensor and rfid applications. IEEE Trans Circuits Syst I Regul Pap 54:153–166. doi:10.1109/TCSI.2006.887974
30. Facen A, Boni A (2006) Power supply generation in CMOS passive UHF RFID tags. In: Research in microelectronics and electronics 2006, Ph D, pp 33–36
31. Ki W-H, Lu Y, Su F, Tsui C-Y (2012) Analysis and design strategy of on-chip charge pumps for micro-power energy harvesting applications. In: VLSI-SoC: advanced research for systems on chip. Springer, Berlin/Heidelberg, pp 158–186. doi:10.1007/978-3-642-32770-4_10
32. Lu Y, Dai H, Huang M et al (2017) A wide input range dual-path CMOS rectifier for RF energy harvesting. IEEE Trans Circuits Syst II Exp Briefs 64:166–170. doi:10.1109/TCSII.2016.2554778
33. Umeda T, Yoshida H, Sekine S, Fujita Y, Suzuki T, Otaka S (2006) A 950-MHz rectifier circuit for sensor network tags with 10-m distance. IEEE J Solid State Circuits 41:35–41. doi:10.1109/JSSC.2005.858620
34. Nakamoto H, Yamazaki D, Yamamoto T et al (2007) A passive UHF RF identification CMOS tag IC using ferroelectric RAM in 0.35-μm technology. IEEE J Solid State Circuits 42:101–110. doi:10.1109/JSSC.2006.886523
35. Le T, Mayaram K, Fiez T (2008) Efficient far-field radio frequency energy harvesting for passively powered sensor networks. IEEE J Solid State Circuits 43:1287–1302. doi:10.1109/JSSC.2008.920318
36. Mounaim F, Sawan M (2011) Integrated high-voltage inductive power and data-recovery front end dedicated to implantable devices. IEEE Trans Biomed Circuits Syst 5:283–291. doi:10.1109/TBCAS.2010.2103558
37. Papotto G, Carrara F, Palmisano G (2011) A 90-nm CMOS threshold-compensated RF energy harvester. IEEE J Solid State Circuits 46:1985–1997. doi:10.1109/JSSC.2011.2157010

38. Xia L, Cheng J, Glover NE, Chiang P (2014) 0.56 V, −20 dBm RF-powered, multi-node wireless body area network system-on-a-chip with harvesting-efficiency tracking loop. IEEE J Solid State Circuits 49:1345–1355. doi:10.1109/JSSC.2014.2305074
39. J-T W, Chang K-L (1998) MOS charge pumps for low-voltage operation. IEEE J Solid State Circuits 33:592–597. doi:10.1109/4.663564
40. Su F, Ki W-H (2008) An integrated reconfigurable SC power converter with hybrid gate control scheme for mobile display driver applications. In: 2008 IEEE Asian solid-state circuits conference (A-SSCC), pp 169–172
41. Oh S, Wentzloff DD (2012) A −32dBm sensitivity RF power harvester in 130nm CMOS. In: 2012 IEEE radio frequency integrated circuits symposium, pp 483–486

Chapter 5
Linear Regulators for WPT

Abstract Linear regulator is an area-efficient component for voltage regulation that could achieve excellent power supply ripple rejection. With no switching activities as compared to a switched-inductor or switched-capacitor power converter, it can serve as a well-controlled power source for digital and especially noise-sensitive analog circuits. It is very suitable for miniature and low-power systems, such as wireless power receivers for portable and implantable applications. This chapter starts with the basic topologies of linear regulators followed by control loop design. Various circuit techniques are demonstrated through the design of two fully-integrated examples that were implemented in 65 nm and 28 nm bulk CMOS processes, respectively.

Keywords Wireless power transfer • Low-dropout regulator • Power supply rejection • PSRR • Digital LDO • Transient response

5.1 Introduction

As mentioned in the previous chapters, the simplest way to regulate the output voltage of the wireless power receiver is to add a post-stage linear regulator, as shown in Fig. 5.1. Linear regulators can filter out the input supply noise and provide a clean supply voltage to drive noise sensitive sensor circuits, wireless communication front-end circuits, critical paths in VLSI chips, and simply provide a voltage step-down function in low-cost applications [1, 2].

The operation principle of a linear regulator is to dynamically tune the series or shunt resistance of the regulator, such that the resistive voltage divider, formed by the source resistance and the load resistance, always maintains a desired ratio for the intended output voltage. Figure 5.2 shows the schematics of a typical series regulator and shunt regulator. In this chapter, we define the parameters V_{IN} as the regulator input voltage and V_{OUT} as the regulator output voltage. The series regulator consists of a power transistor (M_P) controlled by an error amplifier (EA) and a voltage reference. To obtain a fixed output voltage, the series linear regulator adjusts the voltage headroom ($V_{IN} - V_{OUT}$) by its power transistor. The shunt regulator also consists of a power transistor (M_N) and an EA, but different

Y. Lu, W.-H. Ki, *CMOS Integrated Circuit Design for Wireless Power Transfer*,
Analog Circuits and Signal Processing, DOI 10.1007/978-981-10-2615-7_5

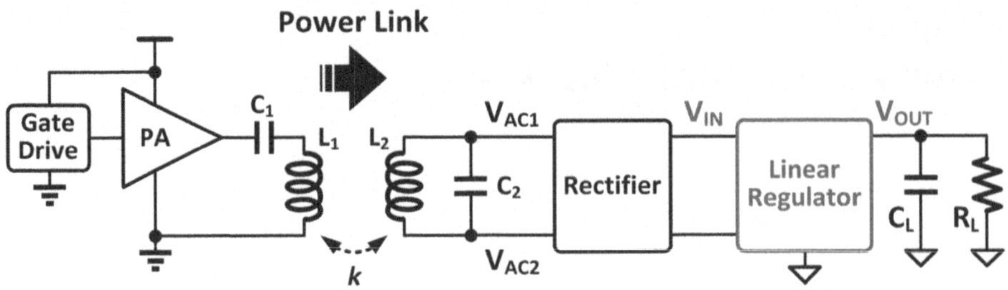

Fig. 5.1 A simple wireless power system with linear regulator for post-stage voltage regulation

from the series regulator M_N consumes an extra current that is parallel to the load. Note that the input source, the rectifier in our case, unavoidably has a finite output resistance, and therefore the extra current drawn by M_N can regulate V_{OUT} accordingly. Since the energy source of the wireless power receiver would change considerably, when the input current is too high, one more important function for the shunt regulator is to bypass the extra energy to ground to prevent the device from over-voltage breakdown.

For both the series and the shunt regulators, they regulate their output voltages by controlling the on-resistance of their respective power transistors, and thus do not generate any output ripples. In addition, a linear regulator can filter out the supply input ripples that come from the previous converter stage. Moreover, a high bandwidth control loop can be relatively easily realized for a linear regulator compared to a switching converter that is usually limited by its switching frequency (especially the inductor-based converter, to be introduced in details in the next chapter). Rather, the bandwidth of a linear regulator is usually limited by the constituent error amplifier and the load it drives, and it could be designed to achieve fast line and load transient responses.

The major drawback of a linear regulator is the unavoidable conduction loss across its power transistor due to the load current that passes through the output voltage drop, and this loss is linearly proportional to $V_{IN} - V_{OUT}$. Note that when V_{IN} gets lower and lower, the output voltage V_{OUT} may not be maintained, and the dropout voltage V_{DO} is defined as the difference between $V_{IN,MIN}$ and V_{OUT}, that is, $V_{DO} = V_{IN,MIN} - V_{OUT}$. In an application if V_{IN} is allowed to be just a few hundred milli-volt or lower above V_{OUT}, then high efficiency could be achieved. However, in such a case, the linear regulator has to have a very low dropout voltage. In fact, in most portable applications, the linear regulators are low-dropout (LDO) regulators. As shown in Fig. 5.2a, the power transistor of an LDO regulator is commonly a P-type transistor because its gate can be driven by a low voltage and it can work in the active region even V_{IN} is very close to V_{OUT}. Nevertheless, the N-type transistor is at times preferred for its intrinsic transient response. Analysis and comparison of these two types are given in the following section.

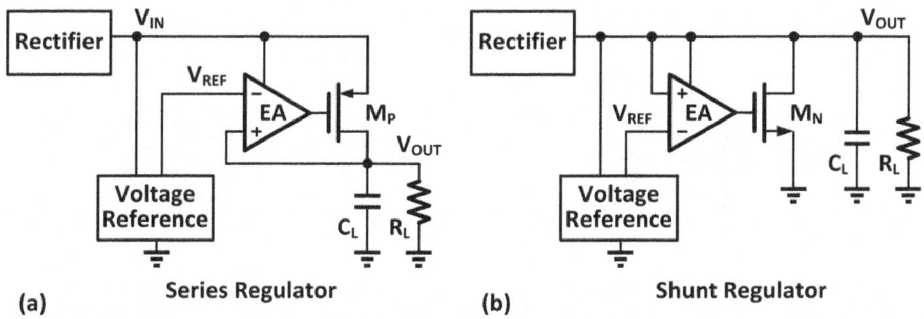

Fig. 5.2 Schematics of (**a**) the series regulator and (**b**) the shunt regulator

5.2 PMOS and NMOS LDO Regulators

The schematic of an LDO regulator with P-type power transistor is shown in Fig. 5.3a and that with N-type power transistor in Fig. 5.3b. Here, we call them the PMOS LDO and the NMOS LDO for short. The power supply of the EA of the PMOS LDO can be directly connected to V_{IN}, because it only needs to output a low voltage for driving $\mathrm{M_{P1}}$. However, the EA of the NMOS LDO should be supplied by a step-up converter (usually a voltage doubler) to generate a sufficiently high voltage to drive $\mathrm{M_{N1}}$. When the output voltage drop is larger than the V_{GS} of $\mathrm{M_{N1}}$, the charge pump may be removed, but then the efficiency would be low. Besides the above mentioned drawbacks of the NMOS LDO, it does have a couple of advantages over the PMOS LDO as discussed below.

The output impedance z_{oP} of the PMOS LDO is

$$z_{\mathrm{oP}} = r_{\mathrm{dsP}} \,||\, \frac{1}{sC_{\mathrm{L}}} \,||\, \frac{1}{A(s)\cdot g_{\mathrm{mP}}}, \tag{5.1}$$

$$z_{\mathrm{oP}} = \frac{r_{\mathrm{dsP}}}{1 + s\cdot r_{\mathrm{dsP}}C_{\mathrm{L}} + A(s)\cdot g_{\mathrm{mP}}r_{\mathrm{dsP}}}, \tag{5.2}$$

where g_{mP} and r_{dsP} are the transconductance and the output resistance of the P-type power transistor, and A(s) is the transfer function of the EA. Similarly, the output impedance z_{oN} of the NMOS LDO is

$$z_{\mathrm{oN}} = r_{\mathrm{dsN}} \,||\, \frac{1}{sC_{\mathrm{L}}} \,||\, \frac{1}{(1+A(s))\cdot g_{\mathrm{mN}}}, \tag{5.3}$$

$$z_{\mathrm{oN}} = \frac{r_{\mathrm{dsN}}}{1 + s\cdot r_{\mathrm{dsN}}C_{\mathrm{L}} + (1+A(s))\cdot g_{\mathrm{mN}}r_{\mathrm{dsN}}}, \tag{5.4}$$

where g_{mN} and r_{dsN} are the transconductance and the output resistance of the N-type power transistor. Note that the multiplicative factors of g_{mP} and g_{mN} are slightly different, and they are A(s) and (1 + A(s)), respectively. As the low-frequency

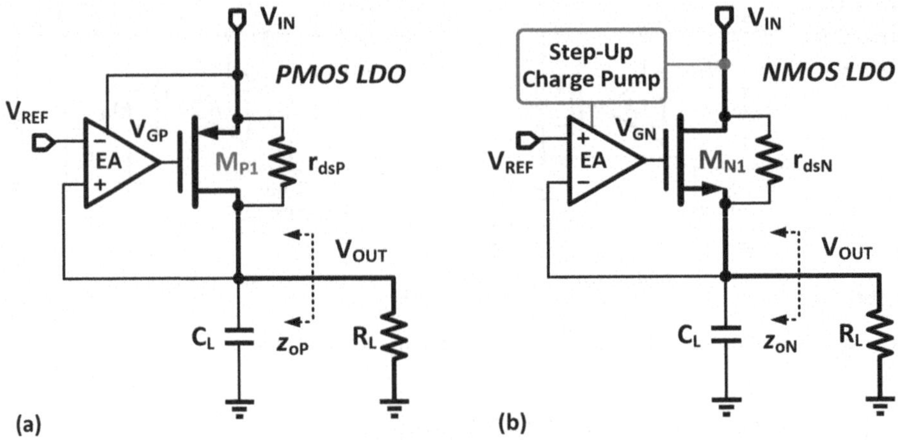

Fig. 5.3 Schematics of low dropout regulators with (**a**) PMOS power transistor or (**b**) NMOS power transistor

(LF) gain is very high ($|A(s)| >> 1$ at LF), the above difference is not important, and we have

$$z_{\mathrm{oP,LF}} \approx 1/A(s) \cdot g_{\mathrm{mP}}, \tag{5.5}$$

$$z_{\mathrm{oN,LF}} \approx 1/A(s) \cdot g_{\mathrm{mN}}. \tag{5.6}$$

However, as the EA has limited bandwidth and cannot respond to the out-of-band high-frequency (HF) signals, hence $A(s) \approx 0$ at HF. Therefore, at high frequency the two output impedances are different, and are given by

$$z_{\mathrm{oP,HF}} \approx 1/sC_{\mathrm{L}}, \tag{5.7}$$

$$z_{\mathrm{oN,HF}} \approx 1/(sC_{\mathrm{L}} + g_{\mathrm{mN}}). \tag{5.8}$$

Obviously, the NMOS LDO has lower output impedance than that of the PMOS LDO at frequencies higher than the unity gain bandwidth of the EA. At such high frequencies, PMOS LDO does not respond to the load changes as fast, and only the load capacitor C_L provides the transient current to the load. For the NMOS LDO, on the other hand, when V_{OUT} drops due to a heavy load step, M_{N1} automatically has a larger V_{GS} and thus provides more current immediately to the load. This is an intrinsic property of a source-follower stage. Therefore, despite a higher driving voltage is required, the NMOS LDO is commonly used for digital loads that generate large and fast load transients and do not have stringent requirement on supply accuracy [3, 4].

In addition, the electrons of an N-type transistor have higher mobility than the holes of a P-type transistor. Hence, the size of an NMOS LDO can be very small, even if the area of the additional step-up charge pump is included. Moreover, in many emerging high-power technologies such as the Gallium Nitride (GaN) and the laterally diffused MOSFET (LDMOS) processes, only the N-type power transistors are available, and circuit designers have the only choice of designing NMOS LDOs.

5.3 Control Loop Design

5.3.1 Dominant Pole Considerations

Capacitors are needed for filtering and compensation but at the same time limit the bandwidth of an analog circuit. For an LDO regulator, the largest capacitors are the output filtering capacitor C_L and the parasitic gate capacitor C_g of the power MOS transistor. Hence, there are at least two low-frequency poles on the left-half-plane (LHP): the pole at the output node $-p_{Out}$, and the pole at the gate of the power MOS $-p_{Gate}$, as sketched in Fig. 5.4 with either $-p_{Out}$ or $-p_{Gate}$ being the dominant pole. For convenience, we will refer to the magnitude of the pole p_X as the pole, while it is understood that the actual pole location is $-p_X$. Now, the pole p_{Out} would shift to a lower frequency when the load resistance increases, and vice versa. Basically, LDO regulators with an off-chip filtering capacitor are designed to be p_{Out} dominant [5–8], while most of the fully-integrated output-capacitor-less LDO regulators have an internal dominant pole [9–13]. Thus, LDO regulators can be classified by the need for an off-chip capacitor or not, or they can be classified by being output-pole dominant or internal-pole dominant [1]. Therefore, there are 4 combinations of which the pros and cons are summarized in Table 5.1 and discussed as follows.

There are many benefits in designing p_{Out} as the dominant pole by using most of the available capacitance (area) at the output node. First of all, a larger output capacitor filters out power supply noise and glitches and serves as a buffer for load-transient current changes, resulting in a smaller ΔV_{OUT}. Second, as shown in Fig. 5.4c and Fig. 5.4d, because the output voltage is well regulated by the control loop at low frequency, and the noise is bypassed to ground by C_L at high frequency, the worst case power supply ripple rejection (PSRR) would occur at medium frequency [14]. Thus, increasing both the output capacitance and the loop bandwidth (that is, the unity-gain frequency, UGF) would improve the PSRR. Third, as the load current decreases, p_{Out} moves to lower and lower frequency, and it is easier to maintain loop stability compared to the internal-pole dominant case. Actually, the zero-load condition had been ignored/omitted in many output-capacitor-less designs, and instead, a minimum load current ($I_{O,Min}$) was required to satisfy the stability requirements. If C_L is reduced to satisfy the stability requirements, the high-frequency PSRR performance will be degraded, and may not be acceptable in many applications. For the WPT receiver, good PSRR at the WPT frequency and its harmonic frequencies (in the 10 MHz or even 100 MHz range) are important to the system.

For the p_{Out} dominant case, pole-zero cancellation is usually used to extend the loop bandwidth and to enhance the stability. The LHP zero may be generated by the

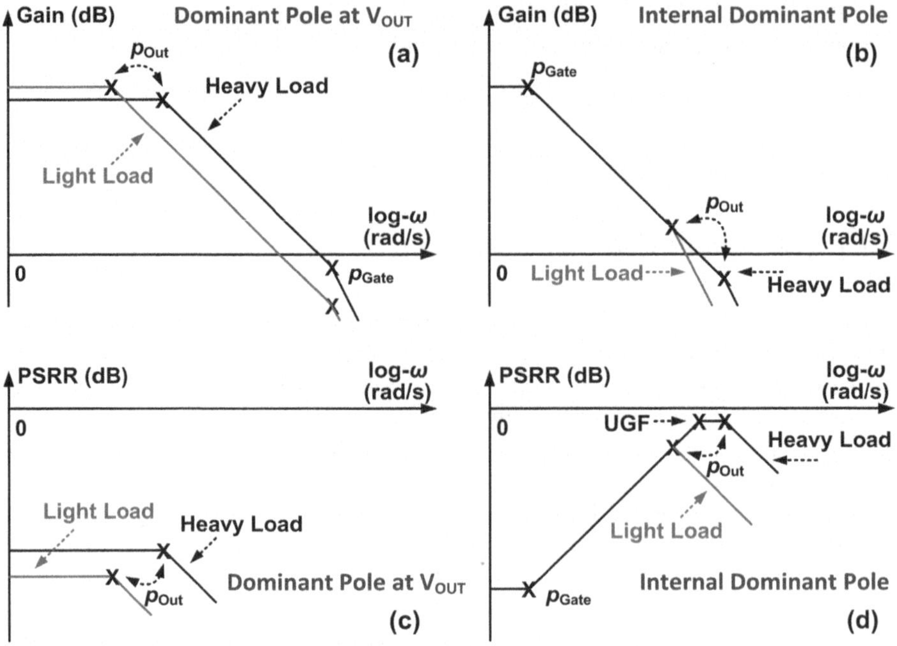

Fig. 5.4 Conceptual frequency response of generic LDO regulators with two LF poles: (**a**) open loop-gain with p_{Out} being its dominant pole, (**b**) open loop-gain with p_{Gate} being its dominant pole, (**c**) PSRR with p_{Out} being its dominant pole, and (**d**) PSRR with p_{Gate} being its dominant pole

Table 5.1 LDO categorization by dominant pole location

Integration	With off-chip cap.		Fully-on-chip	
Dominant pole	p_{Gate}	p_{Out}	p_{Gate}	p_{Out}
Process scaling	No	Yes	No	Yes
Limit on $I_{O,Min}$	Yes	No	Yes	No
UGF	Low	Medium	Medium	High
Transient ΔV_{OUT}	Medium	Small	Large	Medium
PSRR	–	√	×	–
Quiescent current	Low	Medium	Low	High

equivalent series resistance (ESR) of C_L, or by a high-pass feedback network as proposed in [9], or by techniques published in the recent literature. Alternatively, the non-dominant pole p_{Gate} may be pushed to frequencies higher than the UGF by circuit techniques such as adaptive current biasing, g_m boosting, enhanced super source-follower (SSF), etc. [2, 6–8, 15]. The only drawback with the p_{Out} dominant case is that a relatively high quiescent current is needed to push the internal poles to higher frequencies. This requirement can be relaxed by using advanced processes that have lower parasitic capacitance. The transistors will have smaller feature sizes, and the internal poles could be moved to higher frequencies with the same bias current. At the same time, smaller C_L can be used and results in smaller chip area and higher UGF.

To summarize, an LDO regulator with p_{Out} dominant can maintain small V_{OUT} variations and drive large capacitive loads. Also, it can benefit from process scaling that is one of the most desirable characteristics in integrated-circuit design.

5.3.2 Replica Regulator

The replica biasing technique is widely used in source-follower based LDO regulators, for supplying the digital circuits with ultra-fast load-transient responses, or for realizing internal rails for gate-drive buffers. Figure 5.5 shows the schematic of a typical replica regulator that has a replica biasing branch formed by I_B and M_{N1}. The EA only senses V_{MIR} that is the source voltage of M_{N1}, and generates V_G for both M_{N1} and M_{N2}, while M_{N2} supplies the current I_L to the load. Obviously, to make V_{OUT} closely equal to V_{MIR} and consequently V_{REF}, the size ratio of M_{N2} and M_{N1} should be the same as the ratio between I_L and I_B. As the load current I_L changes, V_{OUT} will deviate from the designed value. This structure provides fast load-transient response, but at the same time sacrifices output voltage accuracy that sometimes are acceptable for driving digital loads.

5.3.3 Flipped Voltage Follower

The flipped voltage follower (FVF) [16] based LDO regulator is one of the most popular architectures due to its simplicity and the potential for fast transient response [1, 2, 10, 12, 13]. The schematic of a single-transistor-control LDO based on FVF in [12] is shown in Fig. 5.6 as an example. This circuit can be divided into three parts: the EA, the V_{SET} generation, and the flipped voltage follower. For simplicity, we assume $I_1 = I_2$ and $(W/L)_7 = (W/L)_8$. The mirrored voltage V_{MIR} is controlled by the EA to be equal to V_{REF}, and V_{SET} is generated from V_{MIR} by the diode-connected transistor M_7. Followed by a FVF stage, V_{OUT} is set by V_{SET} through M_8, and it is a mirrored voltage of V_{MIR}. In the FVF stage, M_8 act as a common-gate amplifying stage from V_{OUT} to V_G.

Obviously, there are two low-frequency poles (p_G and p_{Out}) in the FVF loop when a relatively large on-chip C_L is used to handle the large load-transient current. This topology is very difficult (if not impossible) to be stable if p_{Out} is the dominant pole. Another issue associated with this structure is the DC accuracy of V_{OUT}. The offset voltage between V_{REF} and V_{OUT} can be divided into two parts. First, there is an offset between V_{REF} and V_{MIR} that consists of systematic and random offsets of the EA. Second, the mismatches between the voltage mirror (M_7 and M_8) and the bias currents (I_1 and I_2) will generate an offset between V_{MIR} and V_{OUT}. Hence, the FVF-based topology has low immunity to the process, voltage, and temperature

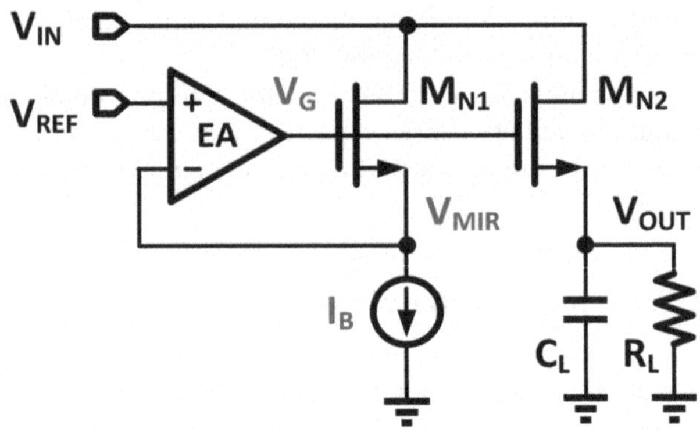

Fig. 5.5 Schematic of a typical replica regulator

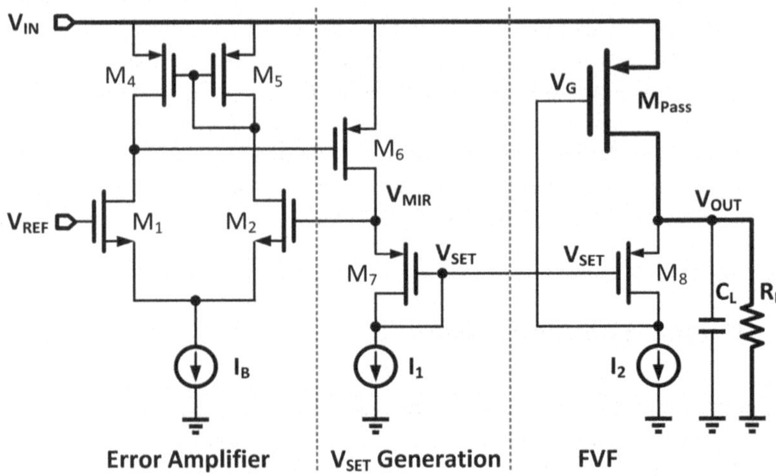

Fig. 5.6 Schematic of the single-transistor-control LDO regulator based on the FVF topology

(PVT) variations. Moreover, the loop gain of the simple FVF is low, which results in poor load regulation, and tens of mV of V_{OUT} variations can be easily observed due to the load current change.

Since the FVF-based LDO regulator is a single-ended topology, for similar dynamic performance, the FVF-based LDO regulator only consumes 50% of the bias current compared to a conventional LDO regulator that uses a differential EA. Although the FVF-based LDO regulator also consists of an auxiliary EA with differential input stage, it is not in the main loop, and only serves as a bias voltage generator and consumes only very low current. Thus, the FVF-based LDO regulator can be more power-efficient. The FVF with folded-cascode gain stage used in [10, 13] can provide higher loop gain and better DC regulation.

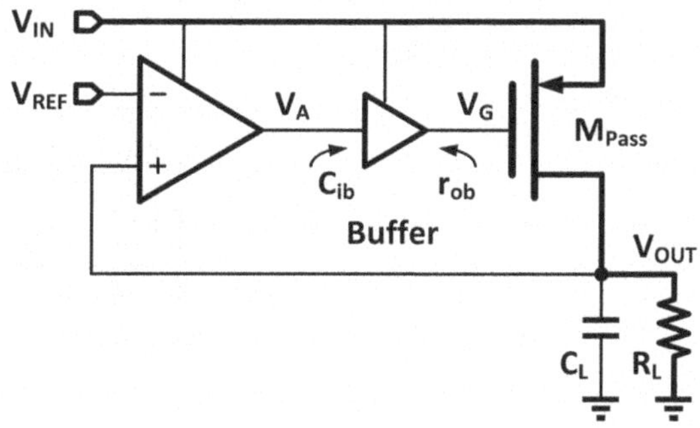

Fig. 5.7 Schematic of the LDO regulator with an impedance attenuation buffer stage

5.3.4 *Impedance Attenuation Buffer*

To effectively push internal poles to higher frequencies, one can split a low-frequency pole into two high-frequency poles. An impedance attenuation buffer can be inserted between the gain stage that has high output impedance and the power stage that has large input capacitance, as shown in Fig. 5.7. The buffer presents low input capacitance to V_A and low output impedance to V_G, therefore, generates two high-frequency poles p_A and p_G. The buffer can simply be a source follower, or a super source follower (SSF) that further reduces the output impedance by employing negative feedback loop [17]. Detailed analysis of the super source follower is given in the design case study in the next section.

5.3.5 *Digitally Controlled LDO Regulator*

Digital low-dropout (D-LDO) regulator has been a popular research topic in recent years for its low-voltage operation and process scalability [18–24]. The basic operation principle is straightforward and can be illustrated by Fig. 5.8. The D-LDO regulator employs one clocked-comparator which can be considered as a 1-bit analog-to-digital converter (ADC), one bi-directional shift-register array and one power transistor array. The $D[1{:}n]$ with unary code controls the number of unit-transistors to be turned on and consequently the total output current. The comparator compares V_{REF} and V_{OUT} in every clock cycle to decide whether the shift register output bits $D[1{:}n]$ shift to the left (add a “1”) or to the right (add a “0”). Obviously, the shift register acts as an integrator in the control loop, and results in a pole at DC. The clocked-comparator can operate at low supply voltages (0.5 V for example) and consumes no static current, while other digital cells can operate at low voltage as well.

The transient response time of the digital LDO regulator is proportional to its clock frequency and the size of each power unit-transistor. Thus, coarse-fine tuning and adaptive clock techniques can be used to improve its transient response without increasing the standby power [22]. In addition, digital codes can be easily

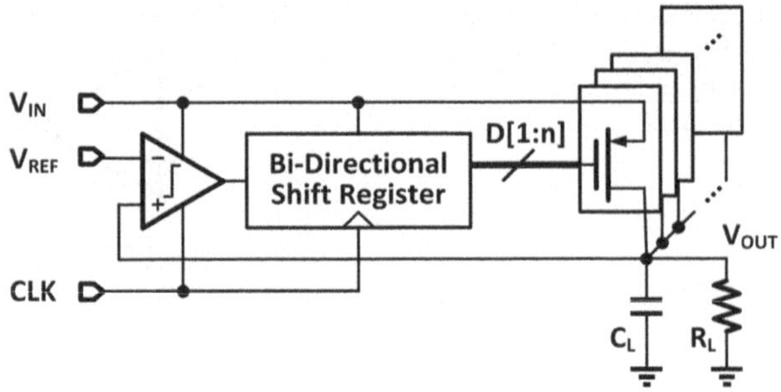

Fig. 5.8 Schematic of a conventional digital LDO regulator

reprogrammed to achieve different step transitions and/or dynamic voltage scaling, which enable convenient system-level reconfigurations. These features of the digital LDO regulator help it to be implemented in low-power low-voltage applications, such as implantable medical devices and wireless sensor nodes.

Similar to the digitally-controlled switching converters, digital LDO regulators also suffer from the problem of limit cycle oscillation (LCO) due to the finite resolution of the output current. When LCO occurs, the code $D[1{:}n]$ will periodically jump between two or multiple consecutive states, and results in undesirable steady-state output voltage ripples. Furthermore, light-load condition results in a larger load resistance R_L, and the same unit-current will generate a larger LCO amplitude. Meanwhile, lower clock frequency results in longer charging and discharging times of C_L, which will also incite a larger LCO amplitude. This problem can be mitigated by using multi-bit ADC, or by adding a feed-forward path with one-bit strength to compensate for the slow main shift-register loop, or by introducing a dead zone to the comparator as discussed in [23, 24].

5.4 Design Case Study

In this section, two LDO regulators, one designed in a 65 nm CMOS process and another in a 28 nm bulk CMOS process will be presented. The basic design guideline for these LDO regulators is to set the dominant pole at the output node by pushing the internal poles to high frequencies, and consequently the LDO regulators enjoy better transient and PSRR performance from process scaling.

5.4.1 Design of Tri-Loop LDO Regulator

To realize a current-efficient LDO regulator with p_{Out} dominant, internal poles should be pushed to high frequencies not only by using large bias current but also with innovative circuit techniques. This case study presents a fully-integrated tri-loop LDO regulator designed in a 65 nm CMOS general purpose (GP) process to achieve ultra-fast transient response and full spectrum (DC to 20 GHz tested) PSRR with limited chip area, current budget and voltage headroom [25].

5.4.1.1 Tri-Loop Design

The transistor-level schematic of the tri-loop LDO regulator is shown in Fig. 5.9, with the signal paths of each loop superimposed on the schematic. Each loop has a different function. Loop-1 is an ultra-fast low-gain loop with the dominant pole p_{Out} at the output, and internal non-dominant poles p_{Gate} and p_{A} are pushed to the GHz range by the impedance attenuation buffer technique. Loop-2 is composed of the EA and the diode-connected M_7 and is a slow loop that generates the voltages of V_{MIR} and V_{SET}. Loop-3 feeds V_{OUT} back to the EA such that the DC accuracy is improved. In other words, Loop-1 is used to deal with the fast load-transient current; Loop-2 generates DC biases; while Loop-3 is used to enhance the V_{OUT} DC accuracy. The output capacitor C_L is 130 pF, the bias current of M_8 is 20 μA, and the buffer consumes another 20 μA at light load (60 μA at heavy load), and all the above assignments help pushing the internal poles to the GHz range.

To increase the DC accuracy of the FVF-based LDO regulator, a third loop is introduced through using a tri-input EA. In conventional architectures, only V_{MIR} is fed forward to generate V_{OUT}, and V_{OUT} is not fed back to the EA. Now, the EA compares V_{REF} with both V_{MIR} and V_{OUT}, and the W/L ratios of the three input transistors M_1, M_2 and M_3 are $(W/L)_1$:$(W/L)_2$:$(W/L)_3$ = 4:1:3 such that V_{OUT} is computed to be

$$\left(V_{\mathrm{REF}} - \frac{1}{4}V_{\mathrm{MIR}} - \frac{3}{4}V_{\mathrm{OUT}}\right) \times \mathrm{A_{EA}} = V_{\mathrm{OUT}} \tag{5.9}$$

$$V_{\mathrm{MIR}} = V_{\mathrm{OUT}} + \Delta V, \tag{5.10}$$

where $\mathrm{A_{EA}}$ is the gain of the EA (including the V_{SET} generation stage), and ΔV is the voltage difference between V_{MIR} and V_{OUT} due to PVT and load variations. By substituting (5.10) into (5.9), and assuming $\mathrm{A_{EA}} >> 1$, we have

$$V_{\mathrm{OUT}} = \frac{\mathrm{A_{EA}}}{1+\mathrm{A_{EA}}}V_{\mathrm{REF}} - \frac{\Delta V \cdot \mathrm{A_{EA}}/4}{1+\mathrm{A_{EA}}} \approx V_{\mathrm{REF}} - \frac{\Delta V}{4}, \tag{5.11}$$

$$\begin{aligned} V_{\mathrm{MIR}} &= \frac{\mathrm{A_{EA}}}{1+\mathrm{A_{EA}}}V_{\mathrm{REF}} + \frac{3\Delta V \cdot \mathrm{A_{EA}}/4}{1+\mathrm{A_{EA}}} + \frac{\Delta V}{1+\mathrm{A_{EA}}}. \\ &\approx V_{\mathrm{REF}} + \frac{3\Delta V}{4} \end{aligned} \tag{5.12}$$

Therefore, V_{OUT} is closer to V_{REF} than V_{MIR} by setting the size ratio of M_2 and M_3 to be 1:3.

Since the EA is not in the high-speed path, the input transistors of the EA and its tail current mirror are implemented with 2.5-V I/O devices for DC gain and

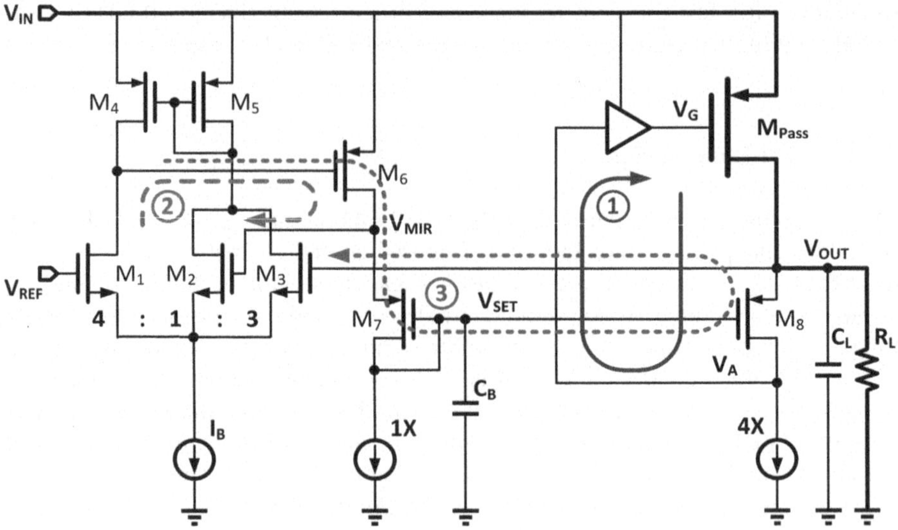

Fig. 5.9 Schematic of the tri-loop LDO regulator

electrostatic discharge (ESD) protection considerations. All on-chip MOS capacitors are I/O devices to avoid gate leakage current if thin-oxide (1.0 V) devices are used. Transistors in the FVF stage are all thin-oxide devices for fast response. To save static current, the ratio of M_7 and M_8, and that of their bias currents, is set to be 1:4, as V_{SET} is in the low-speed path that does not need much current, but V_A is in the high-speed path and needs more current.

5.4.1.2 Buffer Design

The full schematic of the tri-loop LDO regulator including the impedance attenuation buffer is shown in Fig. 5.10. The modified super-source-follower (SSF) buffer used for impedance attenuation consists of M_9 through M_{11}, and three parameters of the buffer are of concern: the input capacitance C_{iB}, the output resistance r_{oB}, and the DC gain A_B. The small-signal model of the buffer is shown in Fig. 5.11. The input capacitance of the circuit can be computed by noting that

$$C_{iB}\Delta V_g = C_{gs}\Delta V_{gs} + C_{gd}\Delta V_{gd}, \tag{5.13}$$

and in the small-signal limit, (5.13) can be rewritten as

$$C_{iB} = \left(1 - v_s/v_g\right) \cdot C_{gs} + \left(1 - v_d/v_g\right) \cdot C_{gd}, \tag{5.14}$$

where C_{gs} and C_{gd} are the gate-to-source and gate-to-drain capacitances of M_9. The voltage gains are calculated as

$$A_B = \frac{v_s}{v_g} = 1 \Big/ \left(1 + \frac{1}{A_1} + \frac{2}{A_1 A_2}\right), \tag{5.15}$$

with

$$A_1 = g_{m9}/g_{ds9}, \tag{5.16}$$

$$A_2 = (g_{m11} + g_{ds12})/g_{dsT}, \tag{5.17}$$

$$g_{dsT} = g_{m10} + g_{ds10} + g_{ds11} + g_{ds13}. \tag{5.18}$$

and

$$\frac{v_d}{v_g} = \frac{v_d}{v_s}\frac{v_s}{v_g} = -\frac{1}{A_2}\frac{v_s}{v_g} = -\frac{A_1}{A_1A_2 + A_2 + 2}. \tag{5.19}$$

Here, A_1 is the intrinsic gain of M_9, and $-A_2$ is the gain from the drain to the source of M_9. Assume A_1 and A_2 to be much larger than 1, then $A_B \approx 1$. Combining (5.14), (5.15) and (5.19), we have

$$\begin{aligned} C_{iB} &= \frac{A_2 + 2}{A_1A_2 + A_2 + 2} C_{gs} + \left(1 + \frac{A_1}{A_1A_2 + A_2 + 2}\right) C_{gd} \\ &\approx \frac{1}{A_1} C_{gs} + \left(1 + \frac{1}{A_2}\right) C_{gd} \approx C_{gd}. \end{aligned} \tag{5.20}$$

Since M_9 operates in the saturation region, most of the channel charge is associated with the source, which means that C_{gs} is much larger than C_{gd}, and hence, C_{iB} is small. The output resistance of the buffer r_{oB} is given by

$$r_{oB} = \frac{1}{A_3(g_{m9} + g_{ds9}) + g_{dsT}}, \tag{5.21}$$

where $A_3 = (g_{m11} + g_{ds12})/(g_{ds12} + g_{ds9})$. To further attenuate r_{oB}, the gain A_3 needs to be increased. The last term g_{dsT} includes g_{m10} and g_{ds10} from M_{10}, and g_{ds11} from M_{11}. Lowering r_{oB} by increasing g_{m10} also increases the pull-up capability of the modified SSF.

A fundamental trade-off of designing the SSF is identified between the DC gain and the frequency response: to satisfy the assumption that $A_1, A_2 >> 1$, the channel length L of M_9 and M_{11} should be long; but to reduce C_{gs} and C_{gd}, L of M_9 and M_{11} should be short. In this design, the minimum L is used for M_9 and M_{11} for speed consideration, and M_{11} operates in the sub- or near-threshold region (in light or full load conditions, respectively) to give a larger g_{m11} to increase A_2 and A_3. Actually, M_9 and M_{11} formed a local negative feedback loop. The gate capacitance of M_{11}, which would generate an additional pole p_D at node V_D, is neglected in the analyses above. This non-dominant pole p_D is located in the GHz range as verified by the following AC simulations and transient measurements.

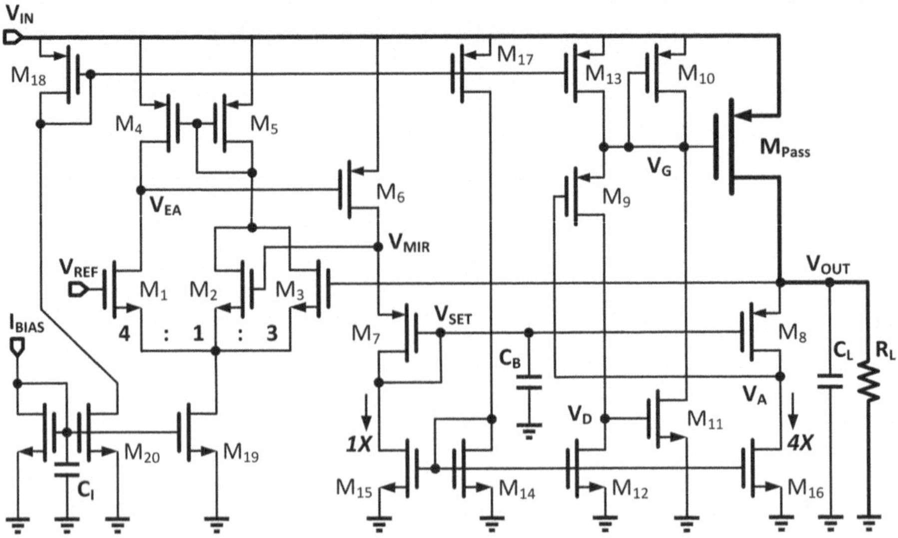

Fig. 5.10 Full transistor-level schematic of the tri-loop LDO regulator

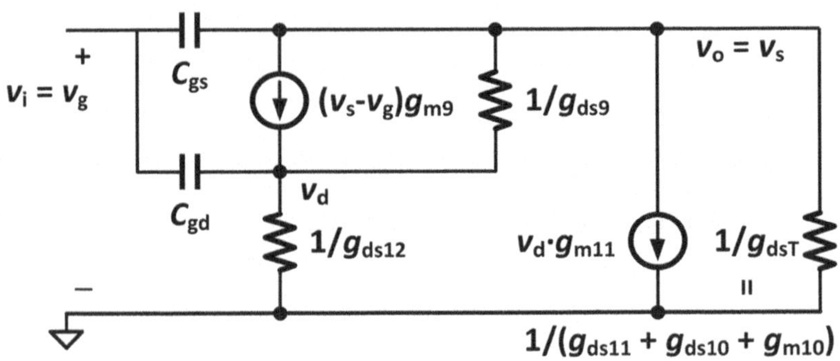

Fig. 5.11 Small-signal model of the super source follower

5.4.1.3 Stability Design

To simulate the frequency response of each loop, three simulation setups are configured and described as follows.

Setup 1 As shown in Fig. 5.12 the signal path of Loop-1 is broken between V_A and the buffer input. The AC small signal is injected to the buffer input and the output is observed at V_A. To isolate the influence from Loop-2 and Loop-3, the path from M_7 to M_8 is also broken. To maintain the DC bias point, a DC voltage V_{SET} is applied to the gate of M_8. And to account for the loading effect, a replica buffer stage is added to V_A to mimic C_{iB}.

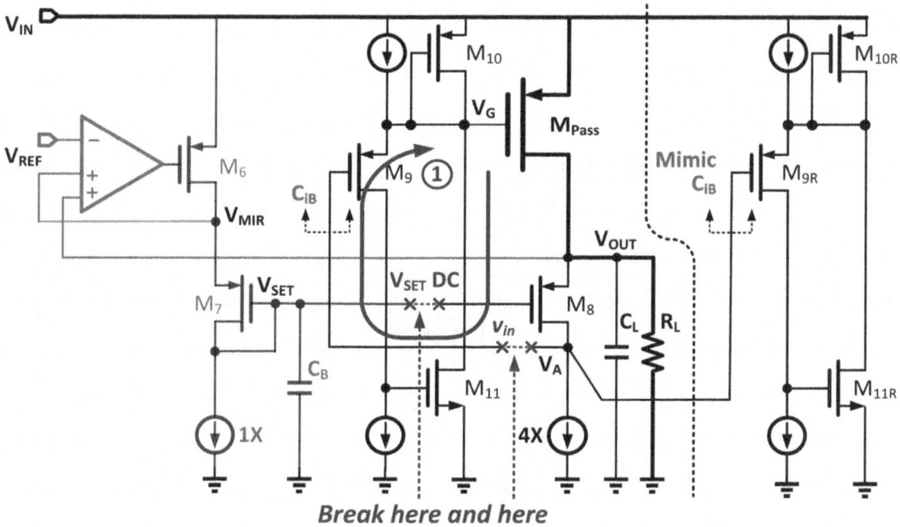

Fig. 5.12 Break Loop-1 with replica buffer connected to V_A to mimic the buffer input capacitance

Setup 2 Loop-2 and Loop-3 are broken from V_{MIR} to M_2 and from V_{OUT} to M_3, respectively, as shown in Fig. 5.13. The AC small signal is injected into the EA through M_2 only. Now, the AC response of Loop-2 can be obtained at V_{MIR}, and the response of Loop-3 can be obtained at V_{OUT}, simultaneously. Since the size ratio of M_2 and M_3 is 1:3, the gain of Loop-3 should be 3 times higher than that of Loop-2. Loop-2 and Loop-3 can be considered together because they both contain the EA in their respective loops.

Simulation results of these two setups are combined in Fig. 5.14, which shows the Bode plots of the three loops at heavy-load condition with $R_L = 100\ \Omega$ and $V_{OUT} = 1.0$ V. Loop-1 has a DC gain of 21 dB and its UGF_1 is 600 MHz, with a phase margin (PM_1) of 60°. Loop-2 has one dominant pole located at V_{SET} and a non-dominant pole located at V_{EA}, and $\mathrm{PM}_2 = 80°$. Loop-3 has two non-dominant poles located at V_{OUT} and V_{EA}, respectively, and PM_3 is only 20°. Nevertheless, the stability of the circuit is determined by the system loop gain, not individual loop gains. A third loop-breaking setup for stability analysis is shown in Fig. 5.15, and described as follows.

Setup 3 Loop-2 and Loop-3 contain the error amplifier, and by breaking the loops between V_{EA} and the gate of M_6 we have

$$v_{ea} = \frac{g_{m2}(r_{o1}||r_{o4})}{1 + sC_{EA}(r_{o1}||r_{o4})}(v_{mir} + 3v_{out}), \tag{5.22}$$

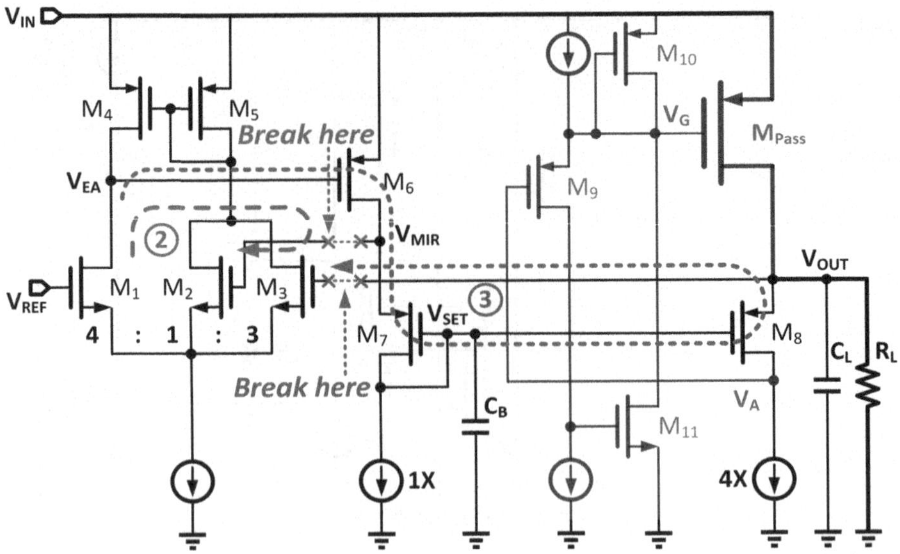

Fig. 5.13 Break Loop-2 and Loop-3 simultaneously for individual analysis

$$\frac{v_{\text{mir}}}{v_{\text{ac}}} \approx \frac{-g_{m6}(r_{o6}||r_{o15})}{1 + sC_B(r_{o6}||r_{o15})}, \tag{5.23}$$

$$\frac{v_{\text{set}}}{v_{\text{mir}}} \approx \frac{1}{1 + 1/A_7(s)}, \tag{5.24}$$

$$\begin{aligned}\frac{v_{\text{out}}}{v_{\text{mir}}} &\approx \frac{1}{1 + 1/A_8 - 1/A_8A_P(s)} \cdot \frac{1}{1 + 1/A_7(s)} \\ &\approx \frac{1}{1 + 1/A_8} \cdot \frac{1}{1 + 1/A_7(s)} = \frac{A_8}{1 + A_8} \cdot \frac{A_7(s)}{1 + A_7(s)},\end{aligned} \tag{5.25}$$

$$A_7(s) = g_{m7}r_{o15}/(1 + sC_Br_{o15}), \tag{5.26}$$

$$A_8 = g_{m8}r_{o8}, \tag{5.27}$$

$$A_P(s) = \frac{g_{mP}(r_{oP}||R_L)}{1 + sC_L(r_{oP}||R_L)}, \tag{5.28}$$

where v_{ac} is the AC signal injected at the gate of M_6, C_{EA} is the parasitic capacitance at the V_{EA} node, and r_{op} is the output resistance of M_{Pass}. The system loop-gain function of the entire tri-loop LDO regulator is given by

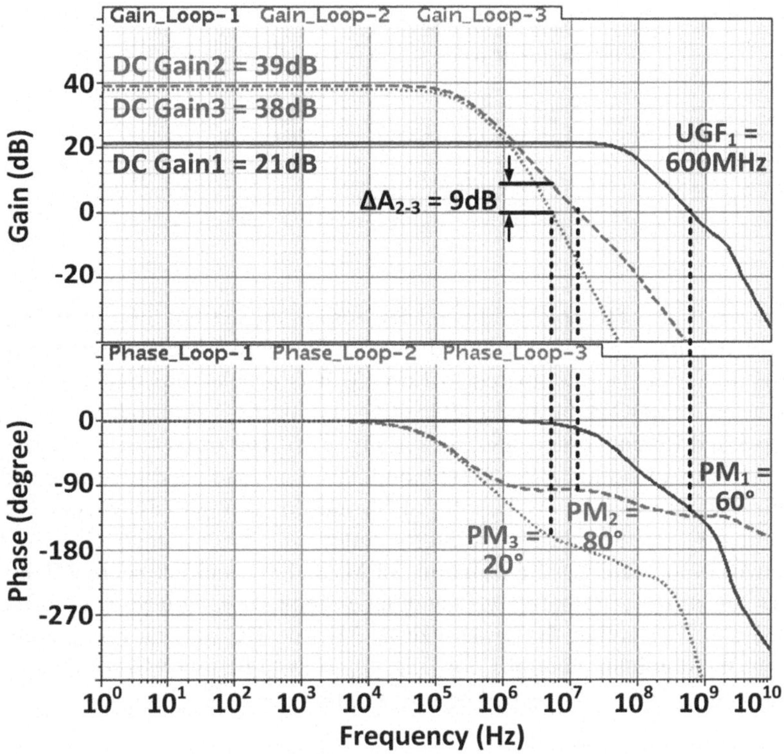

Fig. 5.14 Simulated frequency response of the three loops with $V_{\mathrm{IN}} = 1.2$ V, $V_{\mathrm{OUT}} = 1.0$ V and $R_{\mathrm{L}} = 100\ \Omega$

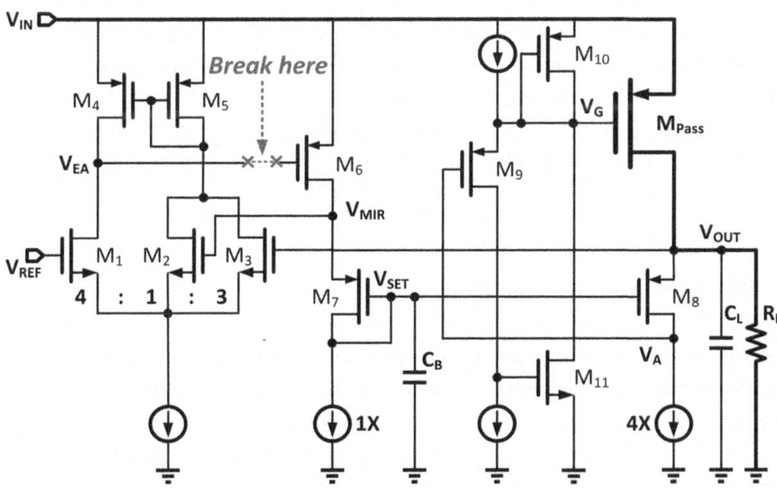

Fig. 5.15 Break the loop at the EA output

$$
\begin{aligned}
T(s) = -\frac{v_{ea}}{v_{ac}} &= \frac{g_{m2}(r_{o1}||r_{o4})g_{m6}(r_{o6}||r_{o15})}{(1+sC_B(r_{o6}||r_{o15}))(1+sC_{EA}(r_{o1}||r_{o4}))} \\
&+ \frac{3g_{m2}(r_{o1}||r_{o4})g_{m6}(r_{o6}||r_{o15})}{(1+sC_B(r_{o6}||r_{o15}))(1+sC_{EA}(r_{o1}||r_{o4}))} \cdot \frac{A_8}{1+A_8} \cdot \\
\frac{A_7(s)}{1+A_7(s)} &\approx \frac{g_{m2}(r_{o1}||r_{o4})g_{m6}(r_{o6}||r_{o15})(1+4g_{m7}r_{o15}+sC_Br_{o15})}{(1+sC_B(r_{o6}||r_{o15}))(1+sC_{EA}(r_{o1}||r_{o4}))(1+g_{m7}r_{o15}+sC_Br_{o15})}.
\end{aligned}
\tag{5.29}
$$

There are three LHP poles and one LHP zero in the system loop-gain function, while the dominant pole is generated by C_B. The zero is generated by Loop-2, which is a shorter path compared to Loop-3. It is a pole-zero tracking pair that makes the entire LDO stable under all loading conditions. The simulated Bode plots of Setup 3 in different corners are given in Fig. 5.16. The worst case phase margins are 68° at 10 mA loading ($R_L = 100\ \Omega$) and 38° at no load condition, respectively.

In this design case, the (W/L) ratio of M_2 and M_3 is aggressively set to be 1:3. This setting is to trade stability margin for better V_{OUT} DC accuracy. To gain more design margin for stability, the weighting of M_2 and M_3 could be set to 2:2 but with lower DC accuracy. Alternatively, in another extreme case, with M_2 (Loop-2) being removed and M_3 having the same size as M_1, the DC accuracy is maximized. However, the dominant pole of Loop-3 at V_{SET} has to be much lower than before, and the settling time of V_{OUT} will be much longer due to a slow Loop-3.

5.4.1.4 Load Regulation

Simulated curves of load regulation are shown in Fig. 5.17, with the size ratio of M_2 and M_3, $(W/L)_2$:$(W/L)_3$, being set to 1:3, 2:2 and 1:0 (no Loop-3), respectively. In the case of no Loop-3, V_{OUT} changed by 34 mV when the load current is changed from 10 μA to 10 mA. For our proposed case of 1:3, V_{OUT} changed by only 11 mV with the same change in load current. DC accuracy is improved by about 3 times by adding Loop-3 without degradation in stability and speed performance. If the ratio of M_2 and M_3 is set to 2:2, V_{OUT} would change by 20 mV.

5.4.1.5 Power Supply Ripple Rejection

PSRR is the most important specification of an LDO regulator designed for noise-sensitive loads. Supply ripples are mainly due to the output voltage ripples from the pre-stage DC-DC converter or AC-DC rectifier, and from the on-chip noise generated by the digital/driver circuits. Ripples generated by the rectifier in the WPT receiver could have harmonics as high as tens of mega-Hertz.

In the first design case, DC gain of Loop-1 has been sacrificed for fast transient response. Increase DC gain of Loop-1 needs additional stages that will introduce undesired LF poles. By setting p_{Out} as the dominant pole, most of the silicon area

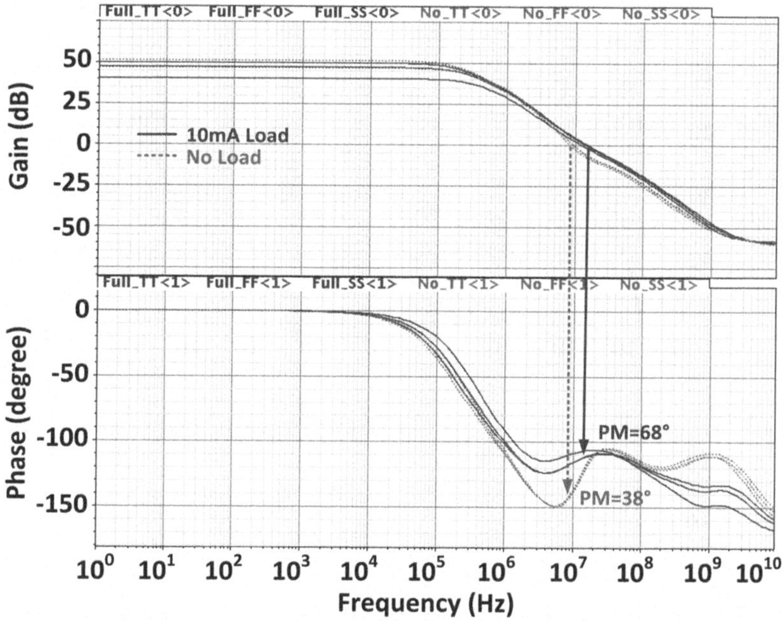

Fig. 5.16 Simulated Bode plot of the tri-loop LDO with $V_{IN} = 1.2$ V and $V_{OUT} = 1.0$ V, at the corners of TT at 25°, SS at 85° and FF at −20°

Fig. 5.17 Load regulation performances with $(W/L)_2$: $(W/L)_3$ being set to 1:3 and 2:2, respectively; and the case without Loop-3

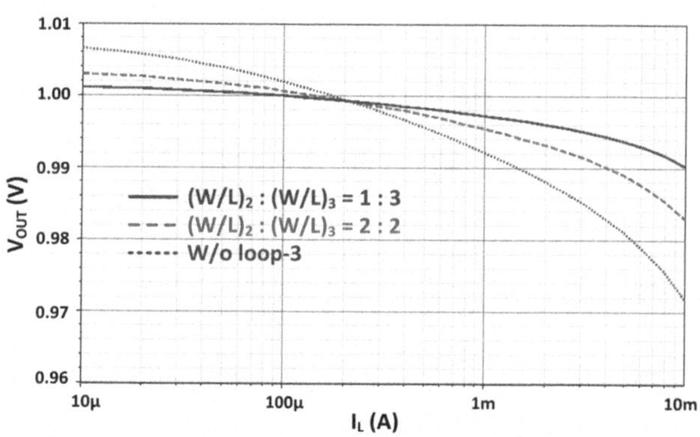

(capacitance) can be effectively used to stabilize V_{OUT} and reject noise from V_{IN}. The simulated PSRR curves of the proposed LDO regulator with full load and no load are shown in Fig. 5.18a; and the PSRR of the tri-loop regulator with and without C_B, and the PSRR of the regulator with only Loop-1 and C_B, are given in Fig. 5.18b, respectively. At medium- and high-frequency ranges, the light-load PSRR is better than the full-load PSRR, because C_L can more effectively bypass the ripple in the VHF (very high frequency, 30 MHz–300 MHz) range to ground when it is in parallel with a larger R_L.

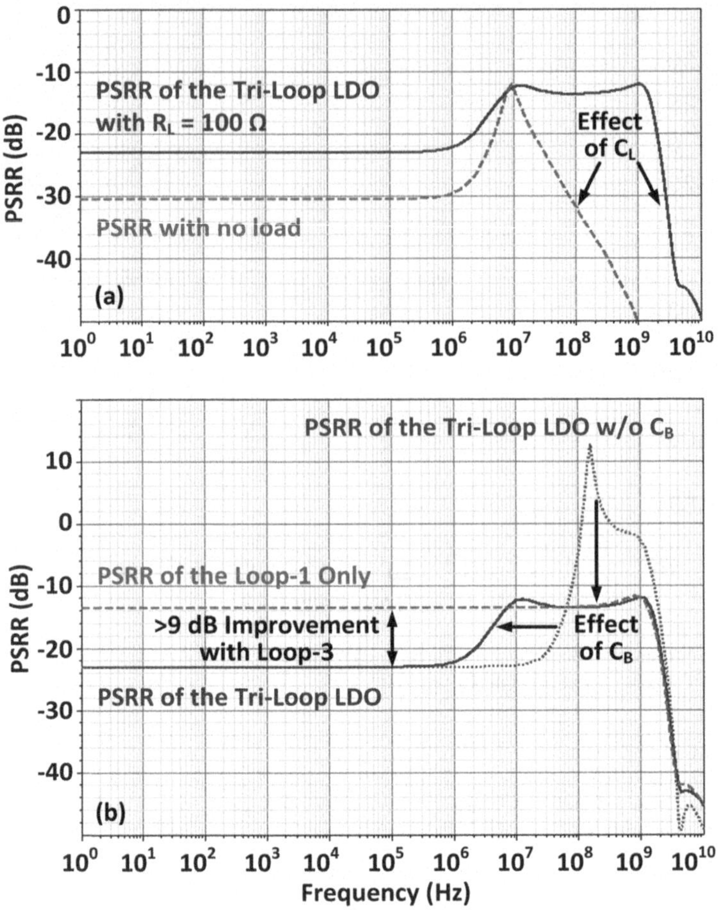

Fig. 5.18 (**a**) Simulated PSRR of the designed LDO regulator; and (**b**) the PSRR of Loop-1 only and the tri-loop regulator with or without C_B, with $V_{IN} = 1.2$ V, $V_{OUT} = 1.0$ V, and $R_L = 100\ \Omega$

The PSRR at low frequencies with Loop-1 only is poor due to its low DC gain. With the designed tri-loop architecture, a 9 dB PSRR improvement is observed at frequencies lower than 1 MHz, compared to the case with Loop-1 only. For the FVF-based structure, V_{OUT} is a strong function of V_{SET}. Therefore, adding a bypass capacitor C_B (about 7 pF in this design) at the V_{SET} node could improve the PSRR by filtering out the ripples that come from V_{MIR} to V_{OUT}. Adding C_B is effective in the medium-frequency range (around 100 MHz–1 GHz), as the stability of Loop-3 is improved with a lower frequency dominant pole at V_{SET}. However, adding C_B will lower the bandwidths of Loop-2 and Loop-3, which are also the PSRR corner frequency of around 1 MHz. The long-channel transistor M_{10} introduces an additional path from V_{IN} to V_G that slightly helps to improve PSRR at high frequencies.

5.4.1.6 Measurement Results

The measurement setup of the LDO regulator with on-chip loading for load-transient measurement is shown in Fig. 5.19. The on-chip R_L is connected in series with the switch S_1 (implemented by a 1.0 V device) driven by an on-chip inverter buffer that achieves load-current edge times T_{Edge} (that is, rise and fall times) of less than 200 ps. The static currents of the chip with S_1 ON and with S_1 OFF are measured as I_{MAX} and I_Q, respectively. The dropout voltage is measured to be 150 mV at I_{MAX} (the worst case). With chip-on-board setup, all the transient waveforms are collected by a pair of 7-GHz differential probes with input impedance of 50 kΩ || 0.32 pF connected to a 4-GHz oscilloscope. Single bond-wire is bonded to each input/output terminal of the prototype. The parasitic *RLC* low-pass filter consists of the 2-nH bond-wire inductance and the input impedance of the probe, of which the cutoff frequency is over 6 GHz. With this setup, ultra-fast transient currents and voltages are generated and measured.

The micrograph of the tri-loop LDO regulator with on-chip loading is shown in Fig. 5.20. The chip area is 260 × 90 μm^2, including 140 pF of on-chip capacitors and the circuit for generating load transients. Fig. 5.21 shows the measured transient response of the output voltage V_{OUT} with on-chip load current change from 0 μA to 10 mA within 200 ps, with zoom-in details of the undershoot and overshoot voltages. With a quiescent current of only 50 μA, the measured undershoot voltage was 43 mV, and V_{OUT} recovered to its steady-state value in 100 ns with the help of Loop-3 regulation. When the load current stepped from 10 mA to 0 μA, the measured overshoot voltage was 82 mV, and V_{OUT} was gradually discharged by the bias current of M_8, and then regulated by Loop-3 to its steady-state value. The well-behaved transient waveforms of V_{OUT} confirmed the stability of the designed tri-loop LDO regulator. To make comparison, a figure-of-merit (FOM) of speed for the LDO regulators is defined in [10] and widely adopted by other researchers. It reads.

$$\text{FOM} = T_R \frac{I_Q}{I_{MAX}} = \frac{C \times \Delta V_{OUT}}{I_{MAX}} \times \frac{I_Q}{I_{MAX}}, \tag{5.30}$$

where I_Q is the quiescent current, and the response time T_R is a function of the total on-chip capacitance C, load-transient glitches of the output voltage ΔV_{OUT} and the maximum load current I_{MAX}. The FOM calculated for this design case is 5.74 ps, with a response time of 1.15 ns. The FOM is expected to be improved further with process scaling, as demonstrated in the following design case with a 28 nm bulk CMOS process. Note that FOM improvement does not necessarily hold for the internal-pole dominant cases, because low loop bandwidth is required by $I_{O,MIN}$ for the LDO regulators to be stable, as discussed in Sect. 5.3.1.

Figure 5.22 shows the measured PSRR of the designed LDO regulator from DC up to 20 GHz. For low frequencies, the PSRR is better than −21 dB; while the worst case occurs at 5 MHz with −12 dB rejection. The PSRR at 1 GHz is −15 dB. For frequencies higher than 2.5 GHz, PSRR would be dominated by the ESR of the

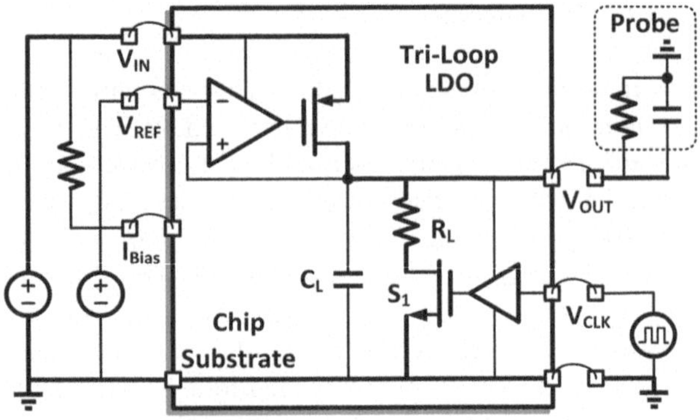

Fig. 5.19 Measurement setup of the LDO regulator with on-chip load

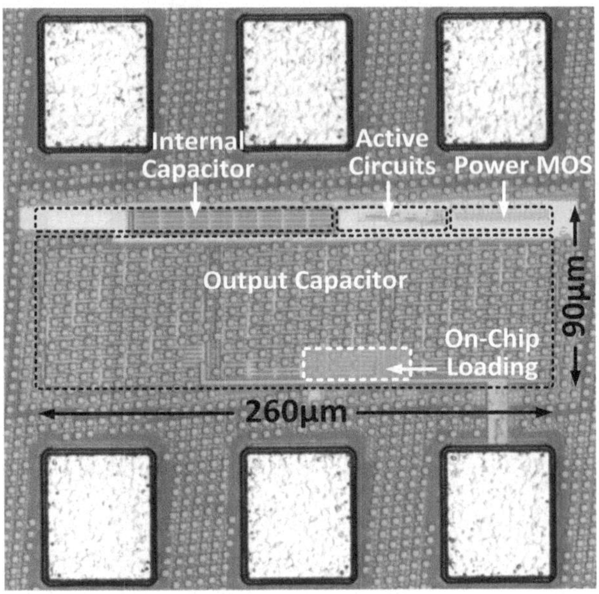

Fig. 5.20 Micrograph of the LDO regulator with on-chip load

filtering capacitors (C_L and C_B). Since the ESR zero is not needed in the proposed architecture, ESRs of the on-chip capacitors are minimized in the layout design by using multiple small capacitors in parallel to achieve good PSRR. Due to the parasitic bond-wire inductance and the substrate-to-PCB resistance, PSRR variations are observed at the VHF region.

Performance comparisons with state-of-the-art LDO regulators are summarized in Table 5.2. Compared to previous designs with ultra-fast transient response [10, 26], response time on the order of nanosecond is achieved by the proposed architecture with much smaller I_Q and C_L, and hence resulting in a very good FOM. Furthermore, full spectrum PSRR characteristic is achieved, while other fully-integrated LDO regulators only give good PSRR at specific frequencies.

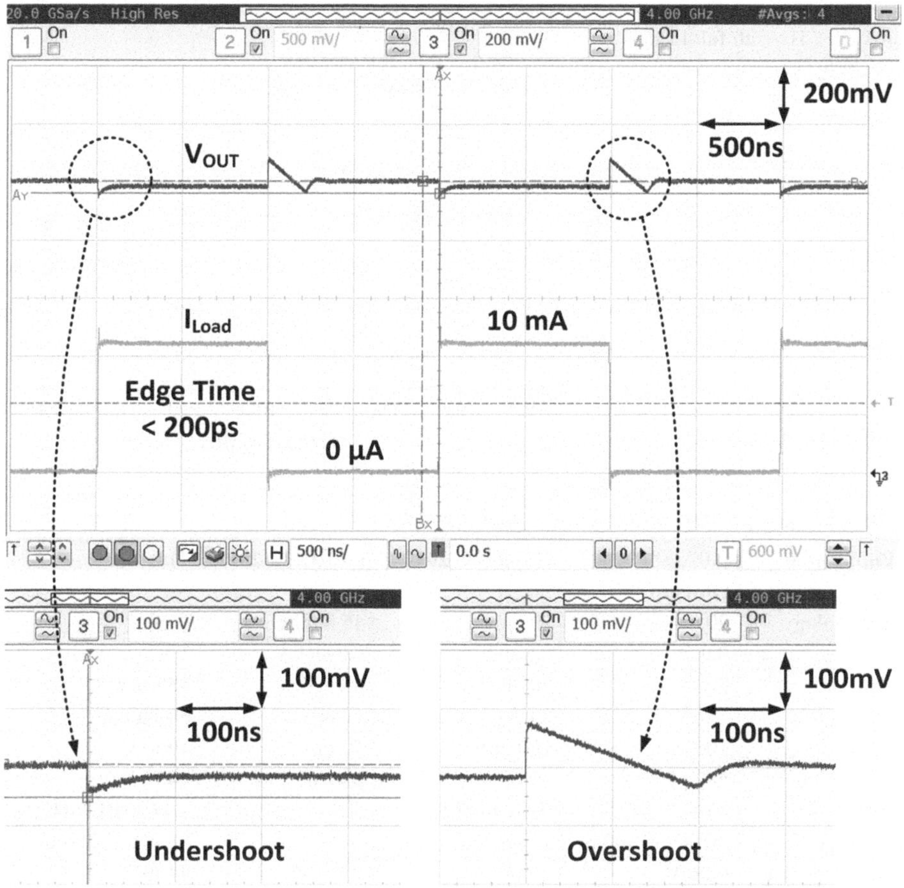

Fig. 5.21 Measured transient response with $V_{IN} = 1.2$ V, $V_{OUT} = 1.0$ V, and on-chip loading change from 0 μA to 10 mA with edge times of 200 ps

5.4.1.7 Conclusions for the Tri-Loop Design

In this case study, a fully-integrated LDO regulator with fast transient response and full spectrum PSRR characteristic is presented. The tri-loop architecture based on the flipped voltage follower and impedance attenuation buffer techniques is designed and verified in a 65 nm CMOS process. With the combined effects of the high-bandwidth Loop-1, C_L and C_B, full-spectrum PSRR is achieved. With the additional Loop-3, V_{OUT} DC accuracy is improved by 3 times compared to the conventional FVF-based LDO regulator. By comparing the performance and design methods of previous non-fully-integrated and fully-integrated LDO regulators, a gap between transient and PSRR performance has been identified and investigated in this research. Of course, higher PSRR in the low and medium frequency ranges will further be improved in the future. As the FOM of this design scales with the

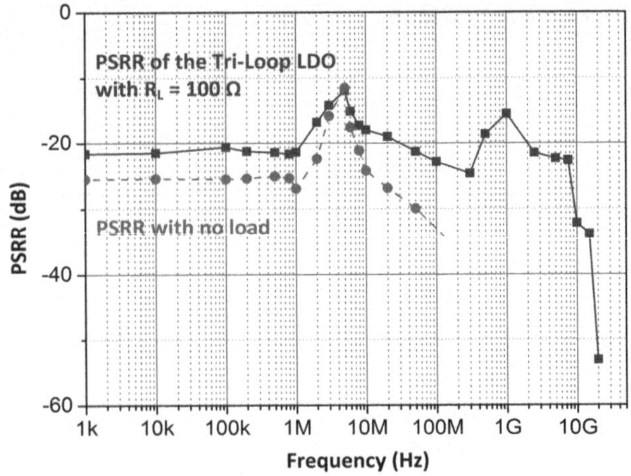

Fig. 5.22 Measured PSRR up to 20 GHz with full load and no-load

Table 5.2 Comparisons for the Tri-loop LDO regulator

Publication	[10] **JSSC 2005**	[13] **JSSC 2010**	[26] **JSSC 2012**	**This work**
C_L	**On-chip**			
Technology	90 nm	90 nm	45 nm SOI	**65 nm**
V_{OUT}	0.9 V	0.5–1 V	0.9–1.1 V	**1 V**
Dropout	300 mV	200 mV	85 mV	**150 mV**
I_Q	6 mA	8 μA	12 mA	**50-90 μA**
I_{MAX}	100 mA	100 mA	42 mA	**10 mA**
Total Cap.	600 pF	50 pF	1.46 nF	**140 pF**
PSRR	N/A	0 dB @1 MHz	N/A	**−15.5 dB @1GHz**
$\Delta V_{OUT}@T_{Edge}$	90 mV @100 ps	114 mV @100 ns	N/A	**82 mV @200 ps**
DC Load Reg.	90 mV	10 mV	3.5 mV	**11 mV**
T_R	0.54 ns	N/A	0.288 ns[a]	**1.15 ns**
FOM	32 ps	N/A	62.4 ps[a]	**5.74 ps**

[a]Based on simulation

process, the proposed architecture will perform better by using more advanced processes.

5.4.2 LDO Regulator with Enhanced Super Source Follower

The cascode FVF topology results in higher DC loop gain and consequently higher UGF, but it is more difficult to make the loop stable, especially when the UGF is in the ultra-high frequency (UHF) band. As shown in Fig. 5.23, with 28 nm process available for the second case study, cascaded buffers are inserted to drive the power

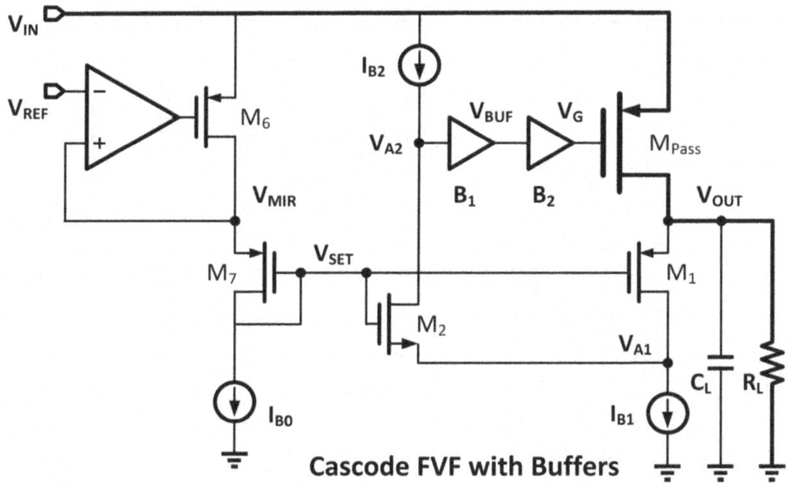

Fig. 5.23 Schematic of the cascode FVF-based LDO regulator with cascaded buffers

transistor M_{Pass}, which only add tiny load capacitance to the node V_{A2}. The first buffer B_1 is simply an NMOS source follower, and the second one B_2 is an enhanced super source follower (E-SSF) [2].

The transistor-level schematic of the cascode FVF-based LDO regulator with E-SSF is shown in Fig. 5.24. Similar to the previous case study, V_{OUT} is a mirrored voltage of V_{MIR}, both of which are one V_{GS} higher than V_{SET}. The function of the left part is to generate the bias voltage V_{SET}. In conventional voltage reference circuit design, V_{REF} is commonly generated by feeding a current through a resistor. When V_{REF} is close to the supply voltage V_{IN}, there is not enough voltage headroom to implement an accurate current source, and the accuracy of V_{REF} will be degraded. To use a lower reference voltage V_{REF}, a resistor ladder of R_1 and R_2 is employed to divide down V_{MIR} and to feed it back to the EA differential input.

The right part of the schematic shows the core circuits of the cascode FVF-based structure with the enhanced super source-follower. M_1 and M_2 serve as two common-gate amplification stages that provide the sufficient DC gain. M_3 is an NMOS source follower that presents low input capacitance to the V_{A2} node, and also shifts V_{A2} down to V_{BUF}, providing more voltage headroom for V_G. To effectively drive M_P, a lower output impedance due to the SSF is needed. In the conventional SSF, only M_4 and M_6 are used, while in the E-SSF, M_5 is inserted between M_4 and M_6. Now, M_4, M_5 and M_6 form a negative feedback loop with higher gain compared to the conventional SSF. The output impedances of the conventional and the enhanced SSF are

$$r_{oB,SSF} \approx \frac{1}{g_{m4} \cdot g_{m6} r_{o4}}, \tag{5.31}$$

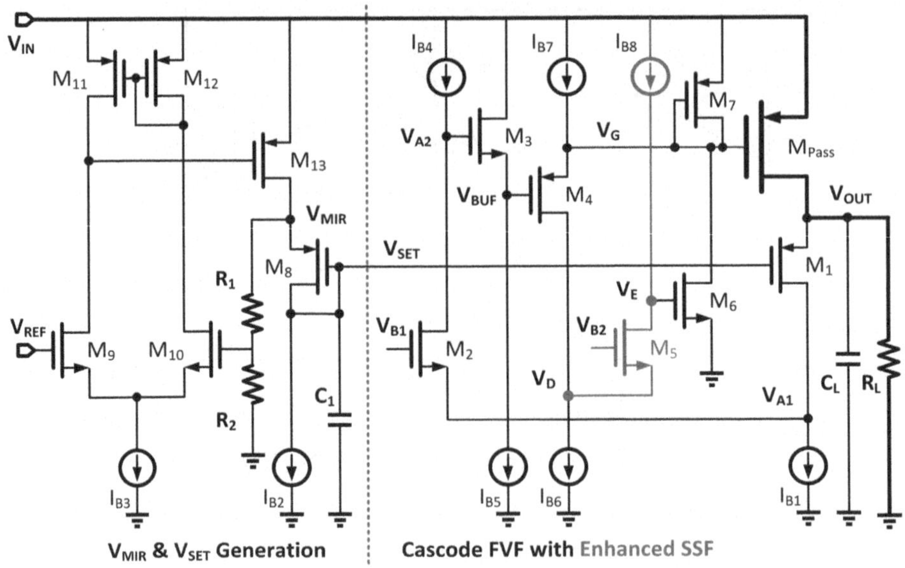

Fig. 5.24 Schematic of the cascode FVF-based LDO regulator with E-SSF

$$r_{\text{oB,E-SSF}} \approx \frac{1}{g_{m4} \cdot g_{m6} r_{o4} \cdot g_{m5} r_{o5}}, \tag{5.32}$$

respectively. The inserted common-gate amplification stage formed by M_5 reduces the output impedance of the SSF buffer by a factor of $g_{m5}r_{o5}$, therefore, providing a larger driving capability.

To reduce the chip area, C_1 and C_L are implemented by low-voltage low-leakage MOS capacitors, with the values of 19 pF and 100 pF, respectively. An additional 1 pF is connected to the gate of the NMOS current mirrors for decoupling. Although M_2 and M_5 need gate bias voltages V_{B1} and V_{B2}, their gates can be connected to V_{MIR} to save additional bias branches. This bias voltage sharing configuration will not affect the frequency characteristics as both gates of M_2 and M_5 have to be referenced to ground.

The LDO regulator is simulated in a 28 nm bulk CMOS process. Figure 5.25 shows the AC responses that include the Bode plots and the PSRR curves, with the load current ranging from 0.1 to 10 mA. The UGF of the cascode FVF loop with the E-SSF is 1.28 GHz with a phase margin of 49°. The low-frequency PSRR is around −27 dB. Benefit from the UHF bandwidth, full-spectrum PSRR is achieved with the worst case of −18.9 dB at 1.55 GHz. The supply noise at above 1 MHz from the left part of Fig. 5.24 is filtered out by C_1. Thus, the PSRR is improved in the 1 to 100 MHz range.

Figue 5.26 shows the transient response when the load current changes from 0.1 mA to 10 mA with the edge time T_{Edge} of 30 ps. The undershoot and overshoot voltages are 26 mV and 21 mV, respectively. The performance summary and

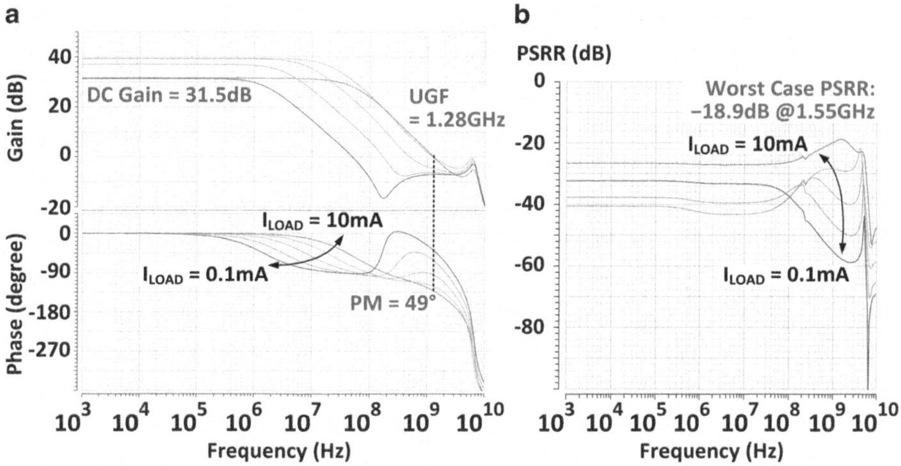

Fig. 5.25 Simulated (**a**) Bode plot and (**b**) PSRR of the proposed LDO with load current ranging from 0.1 mA to 10 mA

comparison with state-of-the-art designs are shown in Table 5.3. Comparing to the tri-loop LDO regulator [1], this work improves the low-frequency PSRR by 6 dB, and also improved the worst-case PSRR by 6 dB. For the load transient response, the change in the output voltage ΔV_{OUT} is reduced by 70%.

5.5 Summary and Discussion

One major difference between a switching power converter and a linear regulator is their nature of energy conversion. The switching power converters use inductors and/or capacitors to store the energy in one phase and then release the energy in the other phase, while linear regulators simply consume or dump away the extra energy. Hence, the efficiency of a series linear regulator is equal to V_{OUT}/V_{IN}, assuming that the quiescent current consumed by the error amplifier is negligible. It is also easier for a linear regulator to achieve a higher bandwidth compared to a switching power converter. However, a linear regulator can only realize the voltage step-down function.

In selecting the type of power transistors, regulators with an N-type power transistor may need a step-up charge pump to provide a higher gate-drive voltage, but it has intrinsic faster load transient response. Regulators with a P-type power transistor are very suitable for the low-dropout design, but may need a larger load capacitor and a more complicated control loop to achieve small output variations. Therefore, the selection of regulator topology should be based on system requirements including but not limited to voltage headroom, current efficiency, speed and noise requirements.

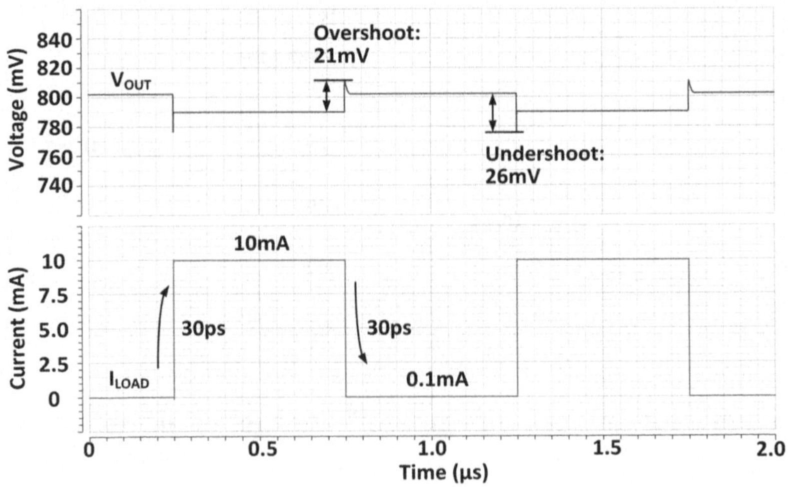

Fig. 5.26 Simulated load transient response of the LDO regulator with $V_{IN} = 1.0$ V, $V_{OUT} = 0.8$ V

Table 5.3 Comparisons of state-of-the-art LDO regulators

Publication	[10] 2005	[27] 2014	[1] 2015	This work
Technology	90 nm	350 nm	65 nm	**28 nm**
V_{OUT}	0.9 V	1.2 V	1 V	**0.8 V**
Dropout	300 mV	600 mV	150 mV	**200 mV**
I_Q	6 mA	44 μA	50–90 μA	**100 μA**
I_{MAX}	100 mA	12 mA	10 mA	**10 mA**
Total Cap.	600 pF	100 pF	140 pF	**120 pF**
Worst Case PSR	N/A	−38 dB @50 MHz	−12 dB @5 MHz	**−18.9 dB @1.55GHz**
$\Delta V_{OUT}@T_{Edge}$	90 mV @100 ps	105 mV @500 ns	82 mV @200 ps	**26 mV @30 ps**
T_R[a]	0.54 ns	N/A	1.15 ns	**312 ps**
FOM[a]	32 ps	N/A	5.74 ps	**3.12 ps**

[a]Based on simulation

As discussed in this chapter, for fast transient response and good PSRR at high frequency, it is highly recommended to design the dominant pole to be at the output node of the regulator, because in such a case most of the available on-chip capacitor (area) is used to filter/attenuate the ripples and disturbances at the output. The design guidelines are verified by two fully-integrated LDO regulator design examples detailed in this chapter. The first design is fabricated in 65 nm CMOS process with a tri-loop architecture, while the second design is implemented in 28 nm bulk CMOS process with enhanced super source follower. Both designs achieved full-spectrum PSRR with limited on-chip capacitors and with a quiescent current of 100 μA.

References

1. Lu Y, Wang Y, Pan Q et al (2015) A fully-integrated low-dropout regulator with full-spectrum power supply rejection. IEEE Trans Circuits Syst I Regul Pap 62:707–716. doi:10.1109/TCSI.2014.2380644
2. Lu Y, Li C, Zhu Y et al (2016) A 312 ps response-time LDO with enhanced super source follower in 28 nm CMOS. Electron Lett 52:1368–1370. doi:10.1049/el.2016.1719
3. Lu Y, Ki W-H, Yue CP (2016) An NMOS-LDO regulated switched-capacitor DC–DC converter with fast-response adaptive-phase digital control. IEEE Trans Power Electron 31:1294–1303. doi:10.1109/TPEL.2015.2420572
4. den Besten GW, Nauta B (1998) Embedded 5 V-to-3.3 V voltage regulator for supplying digital IC's in 3.3 V CMOS technology. IEEE J Solid State Circuits 33:956–962. doi:10.1109/4.701230
5. Rincon-Mora G, Allen PE (1998) A low-voltage, low quiescent current, low drop-out regulator. IEEE J Solid State Circuits 33:36–44. doi:10.1109/4.654935
6. Lam Y-H, Ki W-H (2008) A 0.9V 0.35μm adaptively biased CMOS LDO regulator with fast transient response. In: 2008 IEEE international solid-state circuits conference – (ISSCC), pp 442–626
7. El-Nozahi M, Amer A, Torres J et al (2010) High PSR low drop-out regulator with feed-forward ripple cancellation technique. IEEE J Solid State Circuits 45:565–577. doi:10.1109/JSSC.2009.2039685
8. Ho M, Leung KN, Mak K-L (2010) A low-power fast-transient 90-nm low-dropout regulator with multiple small-gain stages. IEEE J Solid State Circuits 45:2466–2475. doi:10.1109/JSSC.2010.2072611
9. Leung KN, Mok PKT (2003) A capacitor-free CMOS low-dropout regulator with damping-factor-control frequency compensation. IEEE J Solid State Circuits 38:1691–1702. doi:10.1109/JSSC.2003.817256
10. Hazucha P, Karnik T, Bloechel BA et al (2005) Area-efficient linear regulator with ultra-fast load regulation. IEEE J Solid State Circuits 40:933–940. doi:10.1109/JSSC.2004.842831
11. Gupta V, Rincon-Mora GA (2007) A 5mA 0.6μm CMOS miller-compensated LDO regulator with −27dB worst-case power-supply rejection using 60pF of on-chip capacitance. In: 2007 IEEE international solid-state circuits conference – (ISSCC), pp 520–521
12. Man TY, Leung KN, Leung CY et al (2008) Development of single-transistor-control LDO based on flipped voltage follower for SoC. IEEE Trans Circuits Syst I Regul Pap 55:1392–1401. doi:10.1109/TCSI.2008.916568
13. Guo J, Leung KN (2010) A 6-μW chip-area-efficient output-capacitorless LDO in 90-nm CMOS technology. IEEE J Solid State Circuits 45:1896–1905. doi:10.1109/JSSC.2010.2053859
14. Gupta V, Rincon-Mora G., Raha P (2004) Analysis and design of monolithic, high PSR, linear regulators for SoC applications. In: Proceedings of IEEE international SOC conference, pp 311–315
15. Al-Shyoukh M, Lee H, Perez R (2007) A transient-enhanced low-quiescent current low-dropout regulator with buffer impedance attenuation. IEEE J Solid State Circuits 42:1732–1742. doi:10.1109/JSSC.2007.900281
16. Carvajal RG, Ramirez-Angulo J, Lopez-Martin A et al (2005) The flipped voltage follower: a useful cell for low-voltage low-power circuit design. IEEE Trans Circuits Syst I Regul Pap 52:1276–1291. doi:10.1109/TCSI.2005.851387
17. Gray PR, Hurst P, Lewis S, Meyer RG (2009) Analysis and design of analog integrated circuits, 5th edn. Wiley, New York
18. Okuma Y, Ishida K, Ryu Y, et al (2010) 0.5-V input digital LDO with 98.7% current efficiency and 2.7-μA quiescent current in 65nm CMOS. In: 2010 IEEE custom integrated circuits conference (CICC), pp 1–4

19. Chen W-C, Ping S-Y, Huang T-C et al (2014) A switchable digital–analog low-dropout regulator for analog dynamic voltage scaling technique. IEEE J Solid State Circuits 49:740–750. doi:10.1109/JSSC.2013.2297395
20. Gangopadhyay S, Somasekhar D, Tschanz JW, Raychowdhury A (2014) A 32 nm embedded, fully-digital, phase-locked low dropout regulator for fine grained power management in digital circuits. IEEE J Solid State Circuits 49:2684–2693. doi:10.1109/JSSC.2014.2353798
21. Nasir SB, Gangopadhyay S, Raychowdhury A (2016) All-digital low-dropout regulator with adaptive control and reduced dynamic stability for digital load circuits. IEEE Trans Power Electron 31:8293–8302. doi:10.1109/TPEL.2016.2519446
22. Huang M, Lu Y et al (2016) A fully integrated digital LDO with coarse–fine-tuning and burst-mode operation. IEEE Trans Circuits Syst II Exp Briefs 63:683–687. doi:10.1109/TCSII.2016.2530094
23. Huang M, Lu Y et al (2016) Limit cycle oscillation reduction for digital low dropout regulators. IEEE Trans Circuits Syst II Exp Briefs 63:903–907. doi:10.1109/TCSII.2016.2534778
24. Huang M, Lu Y, et al (2016) A digital LDO with transient enhancement and limit cycle oscillation reduction. In: 2016 IEEE Asia Pacific conference on circuits and systems (APCCAS), pp 1–4. doi:10.1109/APCCAS.2016.7803886
25. Lu Y, Ki W-H, Yue CP (2014) 17.11 A 0.65ns-response-time 3.01ps FOM fully-integrated low-dropout regulator with full-spectrum powersupply- rejection for wideband communication systems. In: 2014 IEEE international solid-state circuits conference – (ISSCC), pp 306–307. doi:10.1109/ISSCC.2014.6757446
26. Bulzacchelli JF, Toprak-Deniz Z, Rasmus TM et al (2012) Dual-loop system of distributed microregulators with high DC accuracy, load response time below 500 ps, and 85-mV dropout voltage. IEEE J Solid State Circuits 47:863–874. doi:10.1109/JSSC.2012.2185354
27. Zhan C, Ki W-H (2014) Analysis and design of output-capacitor-free low-dropout regulators with low quiescent current and high power supply rejection. IEEE Trans Circuits Syst I Regul Pap 61:625–636. doi:10.1109/TCSI.2014.2300847

Chapter 6
DC-DC Converters for WPT

Abstract Operation principles of both inductor-based and switched-capacitor DC-DC converters are introduced in this chapter. Different aspects of these two types of DC-DC converters are investigated and compared. In particular, benefits and design obstacles of the time-interleaving multiphase topology are reviewed and discussed. Simulation and measurement results show that the unity-gain bandwidth of a pulse-frequency modulation (PFM) control loop can be designed to be a few times higher than the switching frequency suitable for fast transient applications.

Keywords Wireless power transfer • DC-DC converter • Switched-capacitor • Buck • Boost • Pulse-frequency modulation (PFM)

6.1 Introduction

As mentioned in previous chapters, a DC-DC converter can serve as the power supply of the power amplifier (PA), and to modulate the PA output power by adjusting the converter's output voltage. A DC-DC converter can also be used to regulate the output voltage and/or output current of a wireless power receiver efficiently.

The components available for power conversion are shown in Fig. 6.1. They include capacitors, inductors, transformers, diodes, and also MOSFETs that act as variable resistors or switches. Different from linear regulators that use MOSFETs as variable resistors, as introduced in Chap. 5, DC-DC converters are non-linear voltage regulators that use MOSFETs as switches. An ideal switch has zero resistance when closed and infinite resistance when opened, and hence is lossless whether it is closed or opened. This is one of the two key benefits of using a switching converter over a linear regulator. Another benefit is that a switching converter can be configured to perform voltage step-up, step-down and even polarity inversion, but a linear regulator can only given an output voltage that is lower than the input voltage.

DC-DC converters can be categorized by their energy storage components: the inductor-based DC-DC converter and the switched-capacitor (SC) DC-DC converter. The typical DC-DC converters are shown in Fig. 6.2. Every DC-DC converter contains at least two terminals (T_1 and T_2) that can be assigned as either the input terminal or the output terminal depending on the specific design, one input capacitor and one output capacitor (C_1 and C_2, the functions of which depend

Y. Lu, W.-H. Ki, *CMOS Integrated Circuit Design for Wireless Power Transfer*,
Analog Circuits and Signal Processing, DOI 10.1007/978-981-10-2615-7_6

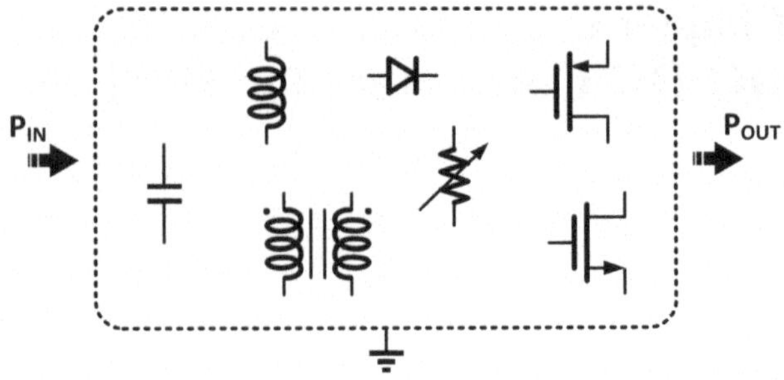

Fig. 6.1 Components available for power conversion

on design), energy storage component(s), and switches. For the inductor-based converter, by switching on and off the switches S_1 and S_2 alternatively, the energy is stored in the magnetic field of the inductor L_1 in one clock phase, and then be delivered to the output in the next clock phase. The output voltage can be controlled by changing the duty cycle of the switches. Based on the relationship between the input and output voltages, a DC-DC converter can be classified as a step-down (buck), step-up (boost) or step-up/down (buck-boost) converter. For the SC converter, the energy is stored in the flying capacitor C_F in one clock phase and be transferred out in the other clock phase. Different from a switched-inductor converter, output voltage regulation of the SC converter is usually achieved by frequency modulation or capacitance modulation [1–4]. An interesting observation of a DC-DC converter is that its direction of energy delivery is reversible. In other words, the input and output terminals are interchangeable, according to the voltage step-up/down requirement of the application. For the converters shown in Fig. 6.2, if the power source is placed on the left such that the power flows from left to right, they are step-down converters; and if the power source is placed on the right and the power flows from right to left, they are step-up converters.

6.2 DC-DC Converter Comparisons

One important performance indicator of a DC-DC converter is efficiency. For applications such as wireless charging of portable devices as discussed in this book, the power range is 1–10 W. Under the same power conversion efficiency, power loss, and thus heat dissipation, increases proportionally as delivered power increases and exacerbates the problem especially for implantable medical devices (IMDs). Energy transfer between an inductor and a capacitor could be lossless; therefore, the ideal efficiency of an inductor-based DC-DC converter could be unity. In practice, efficiency of an inductor-based DC-DC converter is highly dependent on the quality factor of the inductor at the frequency of interest. There are three kinds of losses associated with the power inductor: (1) hysteresis loss due to the difference of magnetic energy being put into the core during the first switching phase and that being extracted from the core during the second switching phase; (2) eddy current loss of the conductive core material and peripherals; and

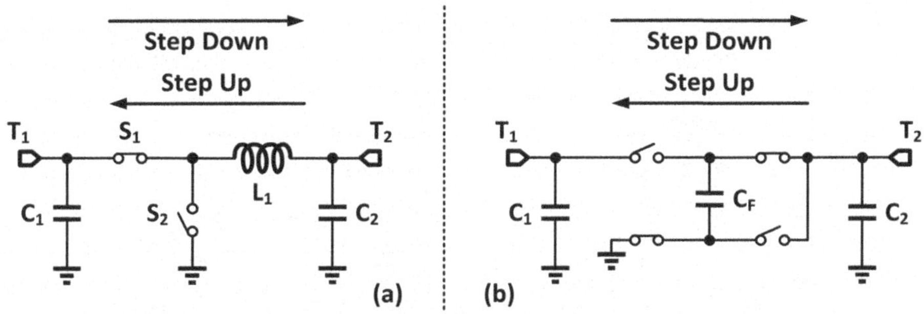

Fig. 6.2 Typical (**a**) inductor-based DC-DC converter and (**b**) switched-capacitor DC-DC converter

(3) conduction loss (ohmic loss, or I^2R loss) caused by the series resistance of the inductor windings. Furthermore, the sizes of high-quality inductors do not shrink as fast as fabrication processes advance, and hence have limited the employment of inductor-based DC-DC converters.

On the other hand, due to the charge redistribution loss [5, 6], the efficiency of an SC DC-DC converter resembles that of a linear regulator, which is equal to the actual output voltage over the ideal no-load output voltage [7]. For a fully on-chip implementation, and especially using a bulk CMOS process, losses due to parasitic capacitors associated with both the top and bottom plates of the flying capacitors degrade the efficiency severely [6]. Note that the parasitic junction capacitance (for example, the N-well to P-substrate junction) of a MOS capacitor could be as high as 7%. Nevertheless, for implantable medical devices the advantage of fully on-chip implementation supersedes the relatively low efficiency, and the SC DC-DC converter is a promising candidate with the other advantages to be discussed next.

Figure 6.3 shows the multiphase/interleaving architectures that are widely adopted for DC-DC conversion to achieve both input current ripple and output voltage ripple reduction. The multiphase topology enables the energy to be delivered to the load in smaller packages, while keeping the switching frequency F_S the same. Consequently, it reduces the ripples without increasing the switching loss. However, a multiphase inductor-based DC-DC converter needs two or more inductors that are hard to be integrated and occupy large chip and/or printed circuit board (PCB) area. It is well known that the inductor-based converter has lower I^2R loss when working in the continuous-conduction mode (CCM). However, CCM operation requires a larger inductor and thus a larger volume. On the contrary, with the advancement of fabrication processes, the on-chip capacitance density has been significantly increased. For example, high power density SC DC-DC converters with good efficiencies have been designed with deep-trench capacitors and/or silicon-on-insulator (SOI) processes [8, 9]. Even in CMOS processes, losses due to parasitic capacitors can be much reduced by appropriate topological considerations and circuit techniques [10–12]. In addition, it is easy for an SC converter to adopt the multiphase architecture with very little power and area overhead [4, 13] and thus achieves ripple reduction, and at the same time, relaxes the requirement

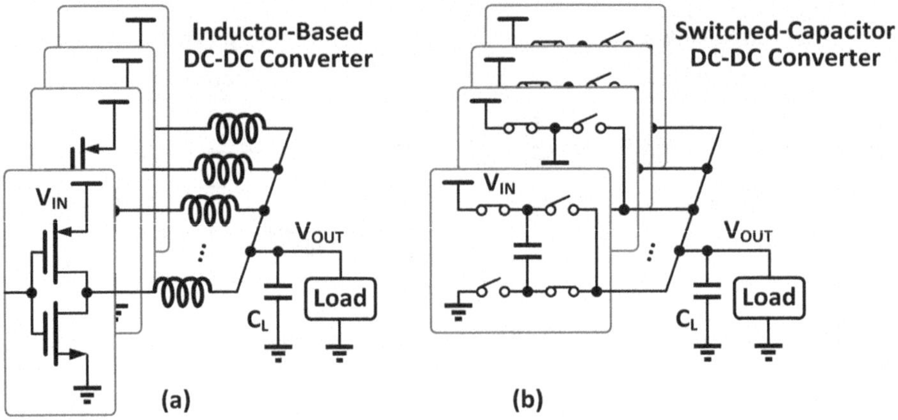

Fig. 6.3 Multiphase (**a**) inductor-based and (**b**) SC DC-DC converters

for both the input and output filtering capacitors. Thus, SC converters are preferred for full on-chip integration in nanometer processes at relatively low power levels [14, 15].

Fully-integrated power converters with fast transient responses are needed for miniaturized devices. A switched-capacitor converter can be relatively easier to achieve faster transient response than its switched-inductor counterpart. Note that an inductor-based DC-DC converter with voltage-mode control contains an LC filter that is of second order, and the control loop needs a complicated compensation scheme (such as a Type III compensator) to maintain system stability while achieving extended bandwidth. In comparison, the switched-capacitor power stage (equivalent to a resistor in series with a capacitor) is only of first order, and pole-zero compensation can be used to extend the bandwidth. Although the power stage of an inductive DC-DC converter with current-mode control can be considered as first-order, the change of the output voltage has a 90° phase delayed compared to the change of the inductor current. The above discussion suggests that for fully on-chip integration, an SC DC-DC converter is more suitable for realizing fast transient response as well. Moreover, because the power stage is of first-order, many design techniques of linear (low dropout, LDO) regulators can be applied to realize bandwidth extension.

6.3 Control Loop Design

6.3.1 Loop Design for Inductor-Based Converters

The voltage conversion ratio (VCR) of the inductor-based DC-DC converter is determined by the duty cycle of the controlling switch, and pulse width modulation

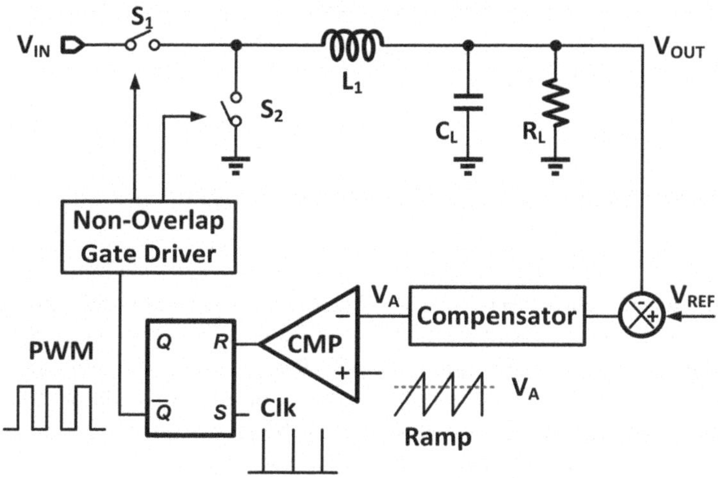

Fig. 6.4 A typical voltage-mode PWM controlled Buck converter

(PWM) can be used in the control loop. Figure 6.4 shows a typical voltage-mode controlled buck converter. The output voltage V_{OUT} is compared with the reference voltage V_{REF} to generate a low-pass filtered voltage V_{A} through the compensator. A sawtooth waveform ramps up at the beginning of each clock cycle that sets the SR latch with a fixed switching frequency F_{S}. When the ramp passes above V_{A}, the SR latch will be reset. The SR latch thus generates the PWM waveform to control the turning on and off of the switches S_1 and S_2.

The switching frequency F_{S} of an inductor-based converter does not necessarily be fixed, and there are control methods with a varying switching frequency such as the constant on-time, constant off-time, pulse skipping, hysteretic control etc., and F_{S} of these control methods would vary with the load.

To maximize the loop bandwidth and achieve fast response, a rule-of-thumb design of the unity-gain frequency (UGF) of the control loop of an inductor-based DC-DC converter is 1/6 of F_{S} [16], because the linearized small-signal model becomes inaccurate when the frequencies of interest approach one-half of F_{S} for a single-phase converter [17].

Studies have confirmed that employing the interleaving topology for the DC-DC converter can extend the theoretic limit of the control-loop bandwidth, and examples can be found for multiphase PWM buck converters [18–21] and for multiphase pulse frequency modulation (PFM) SC converters [22]. As discussed in [19], in a PWM controller, the compensator output voltage V_{A} is sampled by the PWM comparator (which compares V_{A} with the ramp signal), and the perturbation of V_{A} will generate harmonics at the PWM comparator output due to sampling effect. The low-frequency harmonic components that set limit to the control-loop bandwidth can be cancelled by using the multiphase topology. However, the cancelling effect will be degraded due to phase mismatch, and previous works on multiphase buck converters only achieved an UGF slightly higher than F_{S} [20], limited by the

PWM comparator's sampling effect [19] and the small number of phases. Considerations for bandwidth extension of multiphase PFM SC converters are introduced in the following sub-section.

6.3.2 Loop Design for SC DC-DC Converters

For SC DC-DC converters, hysteretic control with single- or multi-boundary is popular due to its fast response and good stability [3, 23]. However, hysteretic or any other ripple-based control methods cannot achieve small output ripple as they need a relatively large output ripple to define the trip points unambiguously.

Figure 6.5 shows a SC converter with voltage-controlled-oscillator (VCO) based PFM control with a load-dependent switching frequency. It has the advantage of high efficiency over a wide load range. The error amplifier (EA) compares V_{OUT} and V_{REF} to tune the VCO frequency through the control voltage V_{CTL}. The centralized current-starved (CS) VCO with a fixed supply voltage (V_{LDO}) is a ring VCO formed by cascading many inverter stages that inherently output multiple interleaving phases for the SC power cells. To reduce the gate drive losses, low-voltage devices are used to implement the switching components; and they are powered up by internal low-voltage domains that prevent device breakdown. Consequently, level-shifters (LSs) are needed to convert the control signals into different voltage domains, and eventually control the switches of the power cells.

However, the conventional PFM control suffers from slow responses and a load-dependent noise spectrum [1, 10]. In a conventional SC converter, the dominant pole of the control loop is located at the V_{CTL} node, and F_S is low at light-load conditions. Thus, the bandwidth is limited and the transient response is slow. To achieve fast load response, Le *et al.* [10] proposed a fast loop triggered by an additional 3.3 GHz clock that bypasses the main integrator loop when the output voltage is lower than a predefined low limit. However, such a high-frequency clock may not be available in many low-power applications, such as the wireless sensor nodes or implantable medical devices. Besides, the reference tracking speed of a converter that implements dynamic voltage scaling (DVS) depends on the

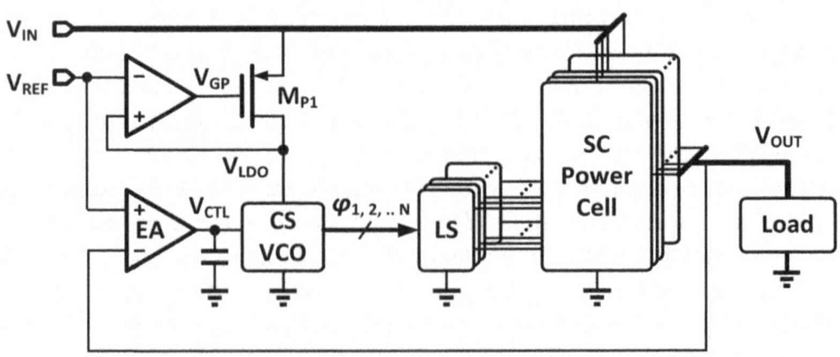

Fig. 6.5 A typical PFM controlled SC DC-DC converter

bandwidth of the main loop, not the large-signal loop. To achieve a high loop bandwidth and small output ripple, a pseudo-continuous-time loop design methodology is introduced next.

As discussed in Chap. 5, there are many benefits of designing the dominant pole of a voltage regulator at its output node. To satisfy the stability requirement, the UGF of the internal-pole-dominated case has to be a few decades lower than the output pole p_O. On the other hand, the UGF of the p_O-dominated case is of course higher than p_O. Thus, the maximum achievable UGF of the p_O-dominated case is higher than that of the internal-pole-dominated case [24]. Furthermore, with a large load capacitor at the output node, the output voltage ripple would be smaller. More importantly, capacitive digital loadings would result in a large load capacitor that gives a low p_O frequency, making the internal-pole-dominated case more difficult to be compensated without sacrificing loop bandwidth.

As shown in Fig. 6.6, centralized clock phases of the conventional SC converter are distributed to the power cells that are scattered over the whole chip through a routing maze, and there could be serious path mismatches that weaken the ripple cancellation effect. Therefore, for the proposed multiphase SC converter, F_S is

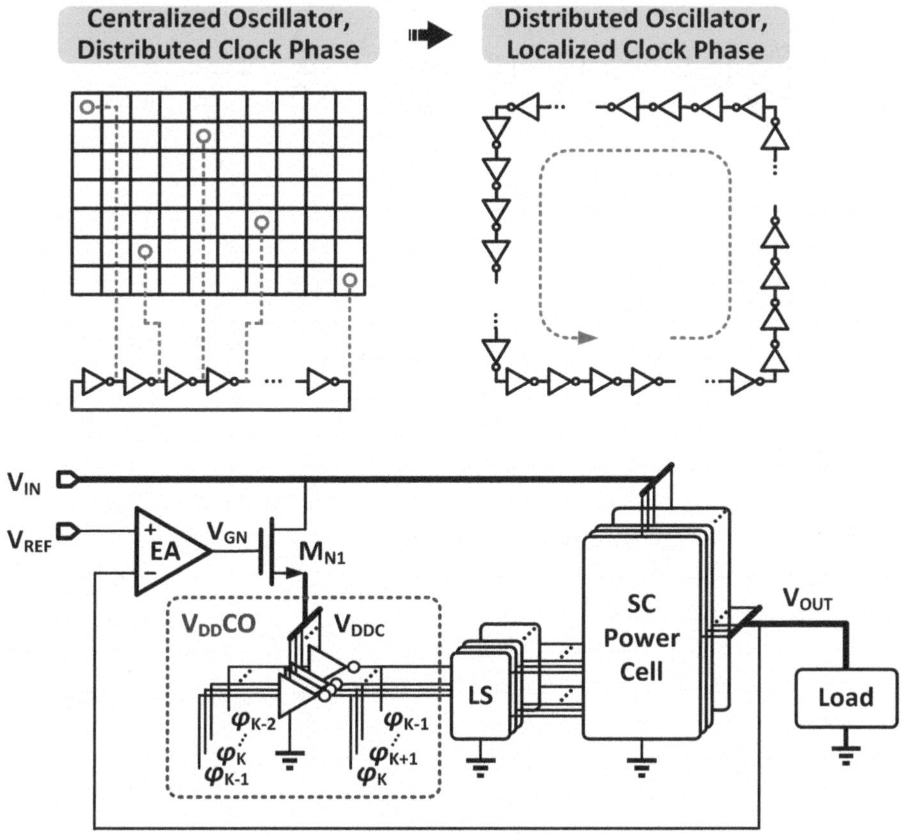

Fig. 6.6 A proposed architecture with distributed V_{DD}-controlled oscillator

determined by the supply voltage of the VCO (V_{DDC}) which is in turn controlled by the EA through comparing V_{OUT} with V_{REF}. The VCO consists of many VCO cells (inverters) that are distributed over the power cells, and the proposed distributed V_{DD}-controlled oscillator (V_{DD}CO) is driven by an NMOS buffer stage. The localized clock phases are the outputs of the distributed inverters among the VCO cells. Note that each current VCO cell is driven by its previous VCO cell, and the phase differences among the cells are significantly reduced. Hence, more interleaving phases can be used. In addition, when the control signal V_{DDC} changes, the output frequency of all the phases changes simultaneously that enables a fast transient response for the PFM control.

Now, the question is: can AC small signals at frequencies higher than F_S pass through the switching power stage, such that the limit is no longer the switching frequency and achieve an UGF that is much higher?

To answer this question, simulations of time-domain AC responses have been carried out for the multiphase SC converter as shown in Fig. 6.7. The SC power stage is a switching stage and the frequencies of interest are close to or higher than

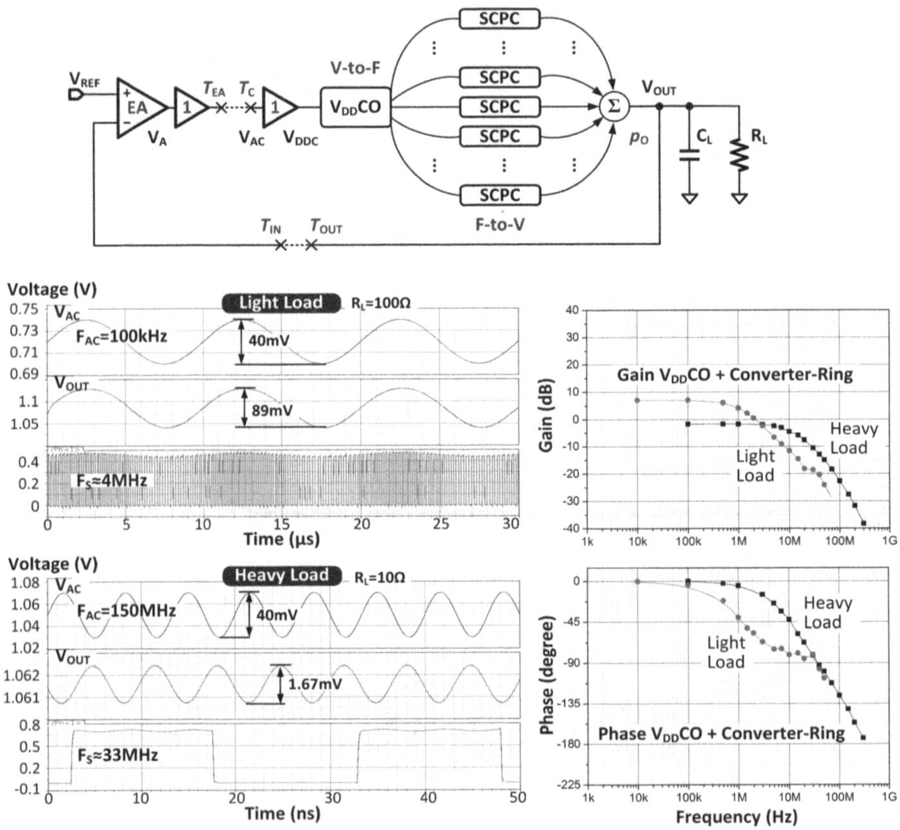

Fig. 6.7 AC simulation of the multiphase SC converter power stage

the switching frequency, hence, the averaged model may not be accurate at high frequencies. Therefore, to verify the validity of the averaged model of the multiphase SC converter, time-domain AC response simulations are conducted, such that the transfer functions of the continuous part and the switching part can be investigated separately, and then be considered together to arrive at the total AC characteristics [22]. Sinusoidal signals with small amplitudes are applied to the buffer before the V_{DD}CO, and the frequency of the applied small signal is swept for computing the frequency response of the power stage. The power stage designed in [22] is employed in this simulation. As shown from the transient waveforms, at light load condition, for example, $R_L = 100\ \Omega$, and F_{OUT} is regulated to around 4 MHz. The small signal of 40 mV$_{PP}$ at 100 kHz injected at V_{AC} is superimposed on V_{OUT} with a magnitude of 89 mV$_{PP}$. At heavy load condition, for example, $R_L = 10\ \Omega$, and F_{OUT} is regulated to around 33 MHz. A small signal of 40 mV$_{PP}$ at 150 MHz was injected at V_{AC}, and was attenuated by the power stage to be 1.67 mV$_{PP}$ and superimposed on V_{OUT}. Similarly, other frequency points were obtained and compiled in the Bode plots (the magnitude plot and the phase plot). Obviously, the SC power stage with the V_{DD}CO has a low-frequency output pole that changes with the load. Therefore, the answer to the question above is yes: the small signals at the frequencies higher than F_S can pass through the switching power stage with multiple interleaving phases.

The VCO here (V_{DD}CO) does not contribute any low-frequency pole to the loop, unlike it does in a phase-locked loop (PLL), because the small-signal information here is not the phase but the frequency. Nevertheless, the multiphase SC power stage can be considered as a phase-integrating block that converts both the phase and the frequency information into V_{OUT} or I_{OUT} (depending on how it is modeled). Here, V_{DDC} is a low-impedance node and the associated pole p_C is located at high frequency, while the output pole p_O becomes the dominant pole. By using the multiphase topology, the control loop can respond to external variations at every fraction of the switching period (T), such that the discrete-time power stage can be considered as a pseudo-continuous-time power stage. Also, different from the previously mentioned PWM case for the inductor-based converter, there is no sampling process for this multiphase SC converter with PFM control. When V_{DDC} changes, the frequency (or the inverter delay) of every phase will be changed simultaneously that ensures pseudo-continuous-time operation. Therefore, the multiphase SC DC-DC converter ring could achieve an UGF a few times higher than F_S, which has been verified on silicon in [22].

6.4 Architectures of DC-DC Conversion

6.4.1 Multiple-Output Converters

Many electronic systems have multiple voltage domains to house different functional blocks, and so does a wireless power transfer system. Different, and sometime isolated, supply voltages are needed for different functional blocks. For example, micro-electrodes use >10 V for neural stimulation; contactless memory uses 15 V for writing and erasing data for flash memory; 3–4 V for battery charging; sub-1 V for low-power microprocessors, etc. Both inductor-based and switched-capacitor DC-DC converters can be reconfigured to generate multiple output voltages by sharing the switches and passive components [25–28]. In particular, by sharing the inductor or capacitor, considerable chip and PCB area could be saved, and thus reduces the device size and cost.

For an inductor-based converter, multiple outputs can be realized by simply delivering the energy stored in the inductor to different outputs in a time-multiplexing fashion [25, 26]. It may also cooperate with a multi-stage rectifier to achieve multi-level operation for output ripple reduction as demonstrated in [26]. To minimize the cross-talk between different outputs, the inductor can be charged once and be discharged to only one of the outputs one at a time [25, 29]. For a simpler controller design, assume that fast comparators are available, the inductor can be charged once and be consecutively discharged to every output one by one [30]. To increase the output current capability, smaller inductance should be used such that the inductor current could ramp up faster, or F_S should be decreased to allow a longer charging time for the inductor [29].

For a SC DC-DC converter, the flying capacitors can be reused by different outputs to realize more VCRs, and consequently to achieve higher system efficiencies over a wider input and/or output range [27]. It is also good to dynamically balance the number of power cells for each outputs according to their load current requirements, such that F_S could be minimized and the efficiency optimized [28].

6.4.2 Layout Strategy for Efficient Power Delivery

The goal of the power distribution system is to deliver the required current across the chip to the load circuits while maintaining the output voltage level for proper operation of the loads. A large DC current I_{DC} will introduce *IR* drop due to the parasitic routing resistance R_P of the V_{DD} network, and the power buses and bond wires will cause V_{DD} variation due to the parasitic inductor L_P during fast load transients. The overall ΔV_{DD} is approximately equal to $I_{DC}R_P + L_P{\cdot}\mathrm{d}i/\mathrm{d}t$ that will cause clock jitter, affect logic delay of the load, and reduce the sensitivity of the analog/RF blocks. Note also that asynchronous circuits are highly dependent on the sequence of logic signals, and are more sensitive to supply variations than synchronous circuits.

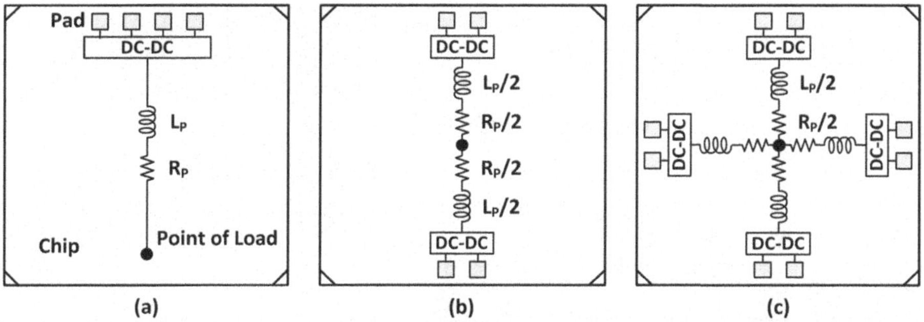

Fig. 6.8 Routing parasitic resistance and inductance of on-chip DC-DC converter(s), supplied from (**a**) one side, (**b**) two sides, and (**c**) all sides

Obviously, as shown in Fig. 6.8a and Fig. 6.8b, by supplying the load current to the center from opposite edges could reduce the worst case R_P and L_P to one fourth of that of supplying the load current from only one edge of the chip. By using DC-DC converters in parallel and supplying the load current from all edges R_P and L_P could further be reduced, as shown in Fig. 6.8c. However, each DC-DC converter needs a control block that takes up area and power. In [22], a multiphase DC-DC converter ring that surrounds the load in the square was implemented. The load can easily get access to the power supply through any point on the edges of the chip. Meanwhile, the in-rush current is reduced by the distributed multiphase configuration and also by using a higher V_{IN} that results in a lower I_{IN}. The converter ring helps reducing the number of power and ground pads while maintaining good noise performance for both V_{DD} and Gnd nodes.

In designing a multiphase SC converter, the issue of phase mismatch has to be addressed through careful layout considerations. For example, if one of the phases is delayed by Δt, then the delay is passed to all subsequent power cells, and a small phase mismatch may result in a large output ripple [22, 31]. As shown in Fig. 6.9, the phase mismatch is modeled as disconnecting the input source from the load for an interval of Δt. When one power cell is not enabled punctually due to phase mismatch, the switch S_O is turned off for the duration of the additional delay. In other words, the delay in one phase will lengthen the overall frequency.

Suppose the DC-DC converter is a multiphase SC converter with the power cells controlled by the distributed ring oscillator. There are two possible architectures for laying out the power cells, as shown in Fig. 6.10: they can either be distributed or centralized. When the power cells are distributed, the process and temperature gradients across the chip and the mismatches in *IR* drop from the supply to each delay cell may result in larger phase mismatches and larger voltage ripples. Meanwhile, the overall switching frequency reflects the averaged process and temperature variations of the entire chip. In addition, with reduced equivalent parasitic inductors and resistors, small output voltage ripples may still be expected. Of course, the selection of the layout architecture depends on system considerations and requires load and power management co-design.

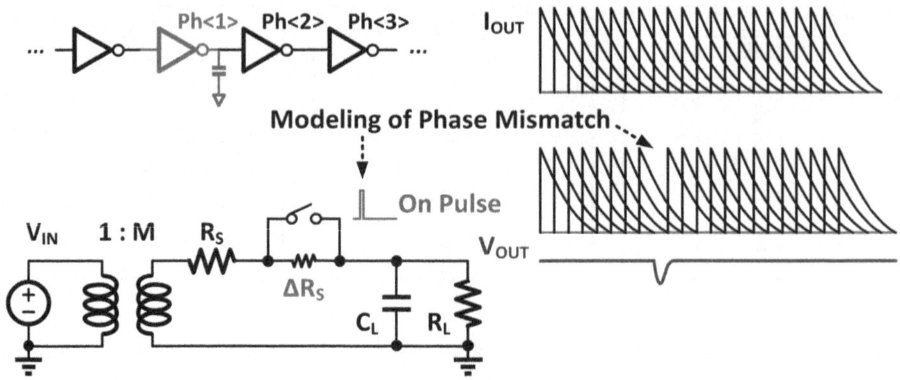

Fig. 6.9 Circuit model of the phase mismatch and the output waveforms of I_{OUT} and V_{OUT}

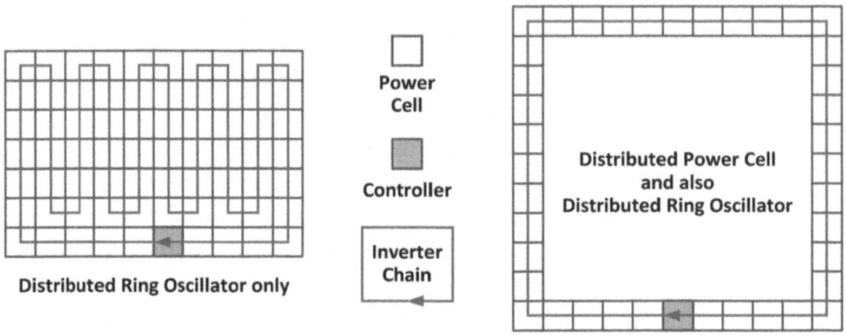

Fig. 6.10 Two possible architectures for the layout of the multi-phase SC converter

6.4.3 Cascade Voltage Regulators

To drive analog/RF loads, the DC-DC switching converter should be cascaded with an LDO regulator, as shown in Fig. 6.11a, and using a PMOS LDO regulator is the common choice. As discussed in the previous chapter, a fully-integrated PMOS LDO regulator may have slow response due to multiple poles, and it is difficult to achieve good power supply ripple rejection (PSRR) at high frequency, especially when the dropout voltage is very low such as 50 mV. Moreover, if the power PMOS has to keep working in the active (saturation) region it has to be very large that makes it hard to achieve fast response and good stability.

The alternative is to use an NMOS LDO regulator, and Fig. 6.11b shows one example that demonstrates a dropout voltage of only 50 mV [2]. In this example, the input V_{IN} is stepped down to V_X through a DC-DC converter and then cascaded by an LDO regulator to generate the output voltage V_{OUT}. The supply voltage of the EA that drives the NMOS power transistor M_{N1} is connected to V_{IN} instead of V_X. Therefore, the supply voltage of the EA is high enough such that the EA output can

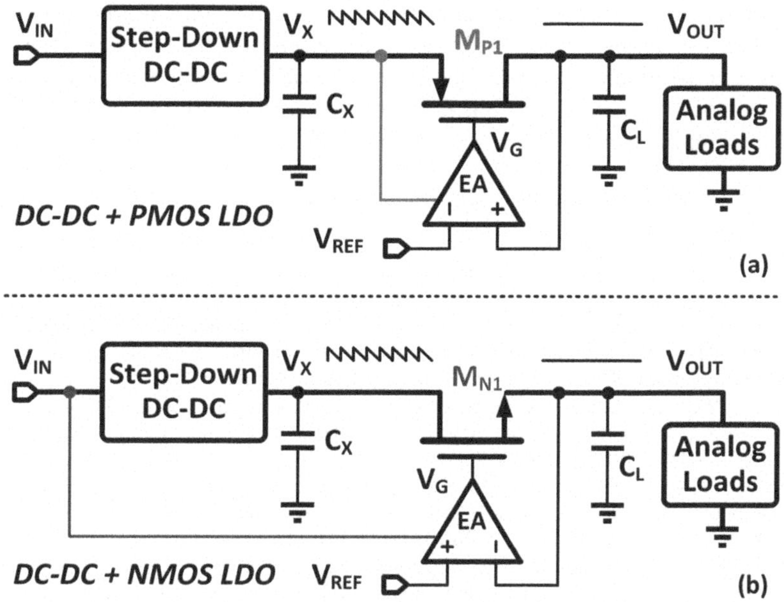

Fig. 6.11 Cascaded voltage regulator stages with (**a**) PMOS LDO regulator of which the EA gets power from V_X, and (**b**) the NMOS LDO regulated topology with EA getting power from V_{IN}

drive the gate of M_{N1} without the need of a step-up charge pump. Note that the output of the DC-DC converter V_X does not need to be tightly regulated, and this is especially favorable for SC converters. Now, only coarse regulation is needed for the SC converter stage, and fine regulation is delegated to the LDO regulator. As such, the SC converter may use digital control with discrete output voltage steps as V_X.

6.5 Summary

For wireless power transfer, and especially for the wireless power receiver, whether using an inductor-based or a switched-capacitor DC-DC converter depends on the power level, device volume, available technology and cost. In general, inductor-based converters are more suitable for high-power applications, and SC converters are more suitable for low-power miniaturized devices. However, the boundary of the preferred power levels shifts with technology, and SC converters are more promising for full integration in advanced processes.

DC-DC converters can adopt multiphase/interleaving technique to achieve input current ripple and output voltage ripple reduction. The output voltage ripple can further be attenuated by cascading a post-stage LDO regulator for noise sensitive functional blocks, such as a signal recording amplifier or an analog-to-digital converter. A major issue associated with the multiphase operation is the phase mismatch problem that has been investigated in this chapter. Careful system-level layout considerations would definitely be helpful to tackle the noise issue. Multiphase operation can also extend the theoretical limit of the control bandwidth of switching converters, and is thus favorable for driving fast transient load such as the power amplifier of a wireless power transmitter.

References

1. Ramadass YK, Fayed AA, Chandrakasan AP (2010) A fully-integrated switched-capacitor step-down DC-DC converter with digital capacitance modulation in 45 nm CMOS. IEEE J Solid State Circuits 45:2557–2565. doi:10.1109/JSSC.2010.2076550
2. Lu Y, Ki W, Yue CP (2016) An NMOS-LDO regulated switched-capacitor DC–DC converter with fast-response adaptive-phase digital control. IEEE Trans Power Electron 31:1294–1303. doi:10.1109/TPEL.2015.2420572
3. Van Breussegem TM, Steyaert MSJ (2011) Monolithic capacitive DC-DC converter with single boundary–multiphase control and voltage domain stacking in 90 nm CMOS. IEEE J Solid State Circuits 46:1715–1727. doi:10.1109/JSSC.2011.2144350
4. Lu Y, Jiang J, Ki W-H, et al (2015) A 123-phase DC-DC converter-ring with fast-DVS for microprocessors. In: 2015 IEEE international solidstate circuits conference – (ISSCC), pp 364–365. doi:10.1109/ISSCC.2015.7063077
5. Ki W-H, Su F, Tsui C (2005) Charge redistribution loss consideration in optimal charge pump design. In: IEEE international symposium on circuits and systems, 2005. ISCAS 2005, vol 2, pp 1895–1898
6. Ki W-H, Lu Y, Su F, Tsui C-Y (2012) Analysis and design strategy of on-chip charge pumps for micro-power energy harvesting applications. In: VLSI-SoC: advanced research for systems on chip. Springer, Berlin/Heidelberg, pp 158–186. doi:10.1007/978-3-642-32770-4_10
7. Seeman MD, Sanders SR (2008) Analysis and optimization of switched-capacitor DC–DC converters. IEEE Trans Power Electron 23:841–851. doi:10.1109/TPEL.2007.915182
8. Le H-P, Sanders SR, Alon E (2011) Design techniques for fully integrated switched-capacitor DC-DC converters. IEEE J Solid State Circuits 46:2120–2131. doi:10.1109/JSSC.2011.2159054
9. Andersen TM, Krismer F, Kolar JW, et al (2015) A feedforward controlled on-chip switched-capacitor voltage regulator delivering 10W in 32nm SOI CMOS. In: 2015 IEEE international solid-state circuits conference – (ISSCC), pp 362–363
10. Le H-P, Crossley J, Sanders SR, Alon E (2013) A sub-ns response fully integrated battery-connected switched-capacitor voltage regulator delivering 0.19W/mm at 73% efficiency. In: 2013 IEEE international solid-state circuits conference – (ISSCC), pp 372–373
11. Jiang J, Lu Y, Huang C, et al (2015) A 2–/3-phase fully integrated switched-capacitor DC-DC converter in bulk CMOS for energy-efficient digital circuits with 14% efficiency improvement. In: 2015 IEEE international solid-state circuits conference – (ISSCC), pp 366–367. doi:10.1109/ISSCC.2015.7063078
12. Butzen N, Steyaert M (2016) A 94.6%-efficiency fully integrated switched-capacitor DC-DC converter in baseline 40nm CMOS using scalable parasitic charge redistribution. In: 2016 IEEE international solid-state circuits conference – (ISSCC), pp 220–221
13. Pique GV (2012) A 41-phase switched-capacitor power converter with 3.8mV output ripple and 81% efficiency in baseline 90nm CMOS. In: 2012 IEEE international solid-state circuits conference – (ISSCC), pp 98–100
14. Villar-Pique G, Bergveld HJ, Alarcon E (2013) Survey and benchmark of fully integrated switching power converters: switched-capacitor versus inductive approach. IEEE Trans Power Electron 28:4156–4167. doi:10.1109/TPEL.2013.2242094

15. Sanders SR, Alon E, Le H-P et al (2013) The road to fully integrated DC–DC conversion via the switched-capacitor approach. IEEE Trans Power Electron 28:4146–4155. doi:10.1109/TPEL.2012.2235084
16. Erickson RW, Dragan M (2001) Fundamentals of power electronics, 2nd edn. Kluwer, Norwell
17. Brown AR, Middlebrook RD (1981) Sampled-data modeling of switching regulators. In: 1981 IEEE power electronics specialists conference, pp 349–369
18. Zhou X, Wong P-L, Xu P et al (2000) Investigation of candidate VRM topologies for future microprocessors. IEEE Trans Power Electron 15:1172–1182. doi:10.1109/63.892832
19. Qiu Y, Yao K, Meng Y, et al (2004) Control-loop bandwidth limitations for multiphase interleaving buck converters. In: 2004 IEEE applied power electronics conference and exposition APEC '04, vol 2, pp 1322–1328
20. Xiao S, Qiu W, Miller G et al (2008) Adaptive modulation control for multiple-phase voltage regulators. IEEE Trans Power Electron 23:495–499. doi:10.1109/TPEL.2007.912947
21. Magallanes FC, Aguglia D, Martins C de A, Viarouge P (2012) Review of design solutions for high performance pulsed power converters. In: Power electronics and motion control conference (EPE/PEMC), 2012, 15th international, pp DS2b.14-1–DS2b.14-6
22. Lu Y, Jiang J, Ki WH (2017) A multiphase switched-capacitor DC-DC converter ring with fast transient response and small ripple. IEEE J Solid State Circuits. doi:10.1109/JSSC.2016.2617315
23. Jiang J, Lu Y, Ki WH (2016) A digitally-controlled 2–/3-phase 6-ratio switched- capacitor DC-DC converter with adaptive ripple reduction and efficiency improvements. In: 2016 42nd European solid-state circuits conference (ESSCIRC), pp 441–444. doi:10.1109/ESSCIRC.2016.7598336
24. Lu Y, Wang Y, Pan Q, et al (2015) A fully-integrated low-dropout regulator with full-spectrum power supply rejection. IEEE Trans Circuits Syst I Regul Pap 62:707–716. doi:10.1109/TCSI.2014.2380644
25. Ma D, Ki W-H, Tsui C, Mok PKT (2003) Single-inductor multiple-output switching converters with time-multiplexing control in discontinuous conduction mode. IEEE J Solid State Circuits 38:89–100. doi:10.1109/JSSC.2002.806279
26. Lu Y, Huang M, Cheng L et al (2017) A dual-output wireless power transfer system with active rectifier and three-level operation. IEEE Trans Power Electron 32:927–930. doi:10.1109/TPEL.2016.2601623
27. Lu Y, Ki W-H, Yue CP (2012) Input-adaptive dual-output power management unit for energy harvesting devices. In: 2012 I.E. 55th international midwest symposium on circuits and systems (MWSCAS), pp 1080–1083. doi:10.1109/MWSCAS.2012.6292211
28. Jiang J, Lu Y, Ki WH, et al (2017) A dual-symmetrical-output switched-capacitor converter with dynamic power cells and minimized cross regulation for application processors in 28nm CMOS. In: 2017 IEEE international solid-state circuits conference – (ISSCC), pp 344–345. doi:10.1109/ISSCC.2017.7870402
29. Jing X, Mok PKT, Lee MC (2011) A wide-load-range constant-charge-auto-hopping control single-inductor-dual-output boost regulator with minimized cross-regulation. IEEE J Solid State Circuits 46:2350–2362. doi:10.1109/JSSC.2011.2162188
30. Le H-P, Chae C-S, Lee K-C et al (2007) A single-inductor switching DC–DC converter with five outputs and ordered power-distributive control. IEEE J Solid State Circuits 42:2706–2714. doi:10.1109/JSSC.2007.908767
31. Somasekhar D, Srinivasan B, Pandya G et al (2010) Multi-Phase 1 GHz voltage doubler charge pump in 32 nm logic process. IEEE J Solid State Circuits 45:751–758. doi:10.1109/JSSC.2010.2042253

Chapter 7
Power Amplifiers for WPT

Abstract Two main choices of the power amplifier (PA) for wireless power transfer (WPT) are Class-D and Class-E PAs. In this chapter, process selection and power loss analysis for high efficiency PAs are discussed first, followed by the discussion on zero-voltage (or zero-current) switching techniques for switching loss reduction. Operation principles and design considerations of both Class-D and Class-E PAs are then discussed. Finally, comparisons are made between these two selections.

Keywords Wireless power transfer • Power amplifier • Class-D • Class-E • Soft-switching • Zero voltage switching (ZVS) • Zero-current switching (ZCS)

7.1 Introduction

DC-AC converters are commonly known as inverters in the power electronics industry, and they are used to convert a DC voltage source into an AC voltage source. Functionally, the non-linear power amplifier (PA) used in analog/RF circuitry resembles a DC-AC converter. It is used to amplify the power of the digitally modulated input signal to be sent out by the transmitter.

In signal processing applications such as audio amplification and wireless communication, signal integrity, signal distortion and harmonics, and signal gain of PA are the main interests. Whereas in wireless power transfer (WPT) applications, efficiency and output power level are the major concerns.

The key feature of non-linear PAs (Class-D and Class-E) as compared to linear PAs (Class-A, Class-AB) is the ideal unity efficiency ($\eta_{ideal} = 100\%$). As mentioned in the previous chapter that discusses DC-DC converters, ideal switches have either zero turn-on voltage or zero turn-off current, and thus do not consume any power. High efficiency is especially important for high power designs, as power loss turns into thermal problem by heating up the system. For example, for a 10 W integrated PA with an efficiency of 90%, 1 W of power loss is converted into heat within a millimeter-sized chip area, and a heat sink that is bulky and costly may be required. If the efficiency drops to 80%, the power loss is doubled, and the size of the heat sink has to be increased accordingly.

Y. Lu, W.-H. Ki, *CMOS Integrated Circuit Design for Wireless Power Transfer*,
Analog Circuits and Signal Processing, DOI 10.1007/978-981-10-2615-7_7

The most commonly used PAs for WPT are Class-D and Class-E PAs [1–9], and they will be discussed in Sects. 7.2 and 7.3, respectively. In the following sub-sections issues related to the PA design and optimization such as integration processes, power loss components and zero-voltage (or zero-current) switching techniques will be introduced. Finally, a brief conclusion is drawn in Sect. 7.4.

7.1.1 Integration Processes

Direct bandgap processes using III/V semiconductor materials such as the Gallium Nitride (GaN) process are popular for high-frequency and high-power applications. The GaN process has high electron mobility, high saturation velocity and high breakdown voltage. PAs fabricated using such processes can achieve higher efficiency and therefore higher power density, and can even operate at higher temperatures [8, 9]. However, the lack of effective oxide layers forbid III/V processes to implement sophisticated digital control circuits and large-scale digital signal processing functions. Therefore, power GaN FETs have to be driven by gate-drivers and controllers implemented in CMOS technologies.

The rightmost drawing of Fig. 7.1 shows the laterally diffused MOSFET (LDMOS). It is an asymmetric power transistor designed to have low turn-on resistance and high drain-to-source and drain-to-gate breakdown voltages. It only needs a small V_{GS} (5 V for example) to turn on, while sustaining very high V_{DS} and V_{GD} (100 V for example). The low doping around the drain terminal results in a diffused depletion layer, labeled as the N− drain extension that can bear a high breakdown voltage. Due to the short effective channel length, small on-resistance can be obtained as well.

Most of the integrated power amplifiers are implemented using the bulk BCD process [10] that contains bipolar junction transistors, CMOS and LDMOS transistors. Bipolar transistors are employed for their high speed and high gain, while CMOS transistors offer high input resistance and low standby power. To increase the output power, either the supply voltage or the output current or both should be increased. To deliver a large current a large transistor is needed; but then it will have large parasitic capacitors at the gate and at the output terminal that result in slow response and large switching loss. To reduce I^2R loss and to attain high efficiency, the PA usually works with a relatively high supply voltage. In addition, to improve the robustness of the integrated power circuits, isolation techniques such as deep-trench isolation (DTI) and silicon-on-insulator (SOI) can be employed, to prevent supply noise and ground noise from coupling among different voltage domains, and to prevent the latch-up problem caused by large in-rush current.

7.1.2 Losses in Switching PAs

Although the theoretical efficiency of a non-linear PA is 100%, in practice, high efficiency is hampered by losses, and the analysis of loss components becomes

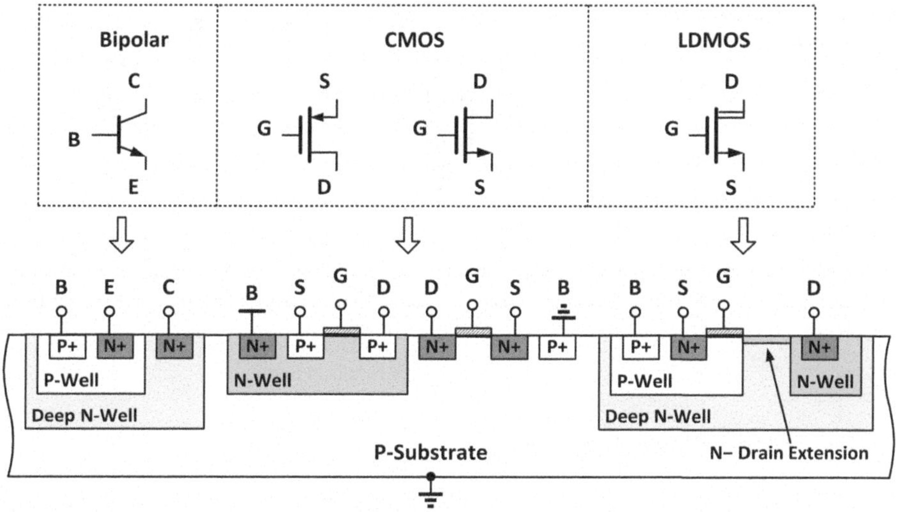

Fig. 7.1 Diagram of combined bipolar, CMOS, and DMOS devices in a BCD process

more important at higher frequencies [11]. Four kinds of loss can be identified and are discussed below.

7.1.2.1 Conduction Loss

The conduction loss, or I^2R loss, is the resistive loss caused by parasitic resistors associated with the power transistors, resonant inductors and capacitors. In addition, the skin effect, which is the phenomenon that high-frequency AC current only conducts near the surface of the conductor and increases the effective resistance at high frequencies, adds to the loss of all conductors, especially for inductors, coils and interconnects.

To reduce conduction loss of interconnects surface mount devices with very low loss at the frequency of interest should be selected. For the bonding pads that conduct large current at high frequencies, multiple parallel bond-wires should be used. Moreover, wide, flat and short connections on chip and circuit board are highly recommended.

7.1.2.2 Turn-On Switching Loss

Figure 7.2 shows an MOS transistor and its model as an ideal switch with a parasitic on-resistor, a parallel body-to-drain diode and a parasitic junction capacitor C_D. Every time the switch is turned on, C_D is discharged to zero with a switching energy loss of $0.5 \times C_D V_{DS}^2$, and the switching power loss is thus $0.5 \times C_D V_{DS}^2 f_s$, where f_s is the switching frequency. In a push-pull type PA such as the Class-D PA, there are

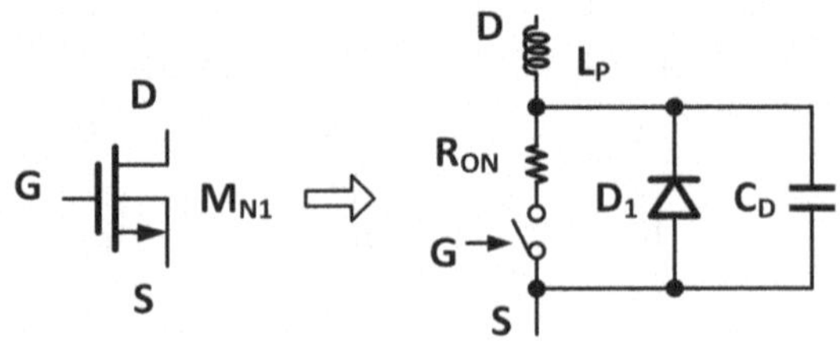

Fig. 7.2 A simple model of an NMOS transistor with parasitics

two junction capacitors at the output node, and there are two switching events in one period. Hence, the switching loss is a few times higher than a single-switch topology.

7.1.2.3 Turn-Off Switching Loss

During the switch turn-off process, the capacitor C_D will be charged up by the power inductor in the PA. At the same time, there is current flowing in the parasitic inductors such as bond-wires and metal traces on the chip and on the board, lumped as L_P, and the amount of energy stored in L_P is $0.5 \times L_P I^2_{OFF}$, where I_{OFF} is the current flowing at the start of the turn-off process. The power loss is proportional to the switching frequency.

Parasitic capacitors are by-products of the devices and are proportional to their sizes that are set by power requirements. Thus, the designer may not have much control. However, inductive power losses can be much reduced by a good circuit layout and PCB layout, and by using an advanced packaging technology.

7.1.2.4 Gate-Drive Loss

Power MOS transistors, due to their large sizes set by power requirement, have large gate capacitances that have to be driven by buffers. The switching power loss of the gate capacitance C_g is $C_g V_{DD}^2 f_S$. Buffers built from cascading inverters with increasing sizes result in large shoot-through current (also known as short circuit current), especially for the later stages. Therefore, non-overlapping logic is also needed for the buffers, as least for the last stage of the buffer.

Power consumption of the controller may be considered as part of the driver's loss. More sophisticated control method can be employed for higher power devices, because larger power budget is available. For a portable application with an output power of 100 mW, designing the controller to consume 1 mW (1% of the output power) is a reasonable starting point. This 1% consumption may be decreased for a larger output power.

7.1.3 Zero Voltage (Current) Switching

As discussed in previous chapters, increasing the switching frequency of power converters can reduce the sizes of capacitors and circuit magnetic, leading to smaller volume and lower cost. However, increasing switching frequency also increases switching loss and thus reduces efficiency. Instead of hard switching, for high-power applications techniques of resonant converters that achieve soft switching have been studied for DC-AC inverters, DC-DC converters, and WPT systems [12–17]. Those techniques make use of resonance of inductor and capacitor to modulate either the voltage or the current waveform across the power switch, so as to achieve either zero-voltage switching (ZVS) or zero-current switching (ZCS) that can reduce the switching loss. As shown in Fig. 7.3, during hard switching, the voltage across and the current through the power transistor are non-zero for the turn-on and/or turn-off transitions. By inciting resonance, either the voltage or the current will be a sinusoidal wave with reduced overlapping. Therefore, switching loss during the turn-on/off transitions can be much reduced.

As discussed in the loss analysis, every time the switch is turned on the voltage across the parasitic capacitor is discharged to zero with a switching loss of $0.5 \times C_D V_{DS}^2$. In the case of ZCS, the MOS transistor has to turn on with a high drain-to-source voltage V_{DS}, and there will be unfavorable switching loss. Consequently, ZCS is not exactly suitable for MOS transistors.

On the other hand, with ZVS the parasitic capacitor discharges its charge and stores it in the resonant inductor, and hence the depletion of charge is not considered as switching loss. After the parasitic capacitor is discharged, the MOS power switch can then be turned on by the gate-drive buffer. Conversely, when the switch is conducting current, it can easily be turned off by the gate signal, and the turn-off losses would not be large. Hence, ZVS is very suitable for MOS transistors, and turning on MOS transistors with a high V_{DS} should be avoided.

7.2 Class-D PA

Figure 7.4 shows a non-linear Class-D push-pull type of power amplifier [18]. CMOS implementation of the power stage of the Class-D PA consists of non-overlapping logic, gate-drive buffer and two power switches. Depending on the choice of process, the two power switches could be a CMOS transistor-pair, or both are N-type LDMOS transistors. The operation of the Class-D PA is similar to that of a buck DC-DC converter. The difference is that the Class-D PA controls the output voltage pulses according to the varying input signal, while the buck converter takes a constant input signal (the reference voltage) and delivers a constant output voltage.

The Class-D PA is traditionally used for audio signal amplification [19], as shown in Fig. 7.4a. The input impedance of a loudspeaker is mainly resistive over

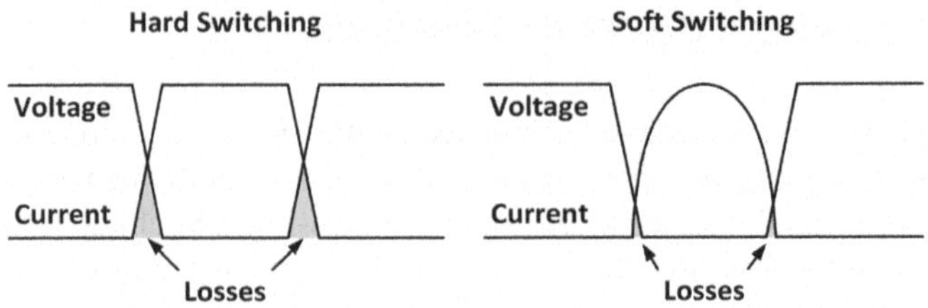

Fig. 7.3 Voltage and current waveforms of hard switching and soft switching scenarios

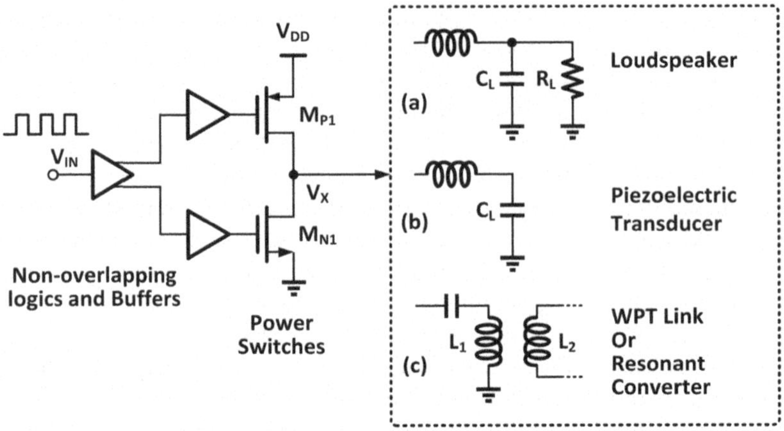

Fig. 7.4 A Class-D type power amplifier driving (**a**) loudspeaker, (**b**) piezoelectric transducer, and (**c**) wireless power transfer link or resonant DC-DC converter

the frequency range of interest, and the output power of the loudspeaker is specified by nominal impedance such as 8 Ω for Hi-Fi loudspeakers and 32 Ω for headphones, for example.

Figure 7.4b shows the driver of a piezoelectric transducer that finds applications in sound generation and actuation. The piezoelectric transducer can be electrically modeled as a capacitive load that requires a high voltage to drive [20], and high-efficiency Class-D designs have been demonstrated with relatively high peak-to-average ratio signals [21].

When driving the load for a WPT link or a resonant DC-DC converter, as shown in Fig. 7.4c, a transformer may be involved. The load along with the resonant tank on the secondary side can be modeled as an equivalent load impedance $Z_{L,EQ}$ reflected back to the primary side (with residual inductive or capacitive component) that will affect the resonant frequency f_R of the primary LC tank. It will be difficult to match the switching frequency f_S of the Class-D PA with f_R exactly in real implementations. Refer to Chap. 3 for more detail discussion.

7.2.1 Operation Conditions

The conceptual voltage and current waveforms of the CMOS Class-D PA driving a resonant load are shown in Fig. 7.5. There are three operation cases. Case 1: the equivalent load seen at the PA output is purely resistive, and hence, $f_S = f_R$. Case 2: the secondary circuit presents a capacitive load with $f_S < f_R$, and the PA output current I_O leads the voltage signal V_X. Case 3: the secondary circuit presents an inductive load with $f_S > f_R$, and the PA output current I_O lags behind the voltage signal V_X.

Assume the transistors are ideal and no dead-time is introduced to the power stage, the node voltage V_X is then a square wave with 50% duty ratio. Thus, V_X has a fundamental sinusoidal component at f_S with odd harmonics. The series-resonant LC tank should have a high loaded quality factor Q_L, and acts like a band-pass filter attenuating the high-order components. Therefore, I_O that flows through the resonant tank can be considered as almost sinusoidal. In Case 1 with $f_S = f_R$, the positive half of I_O is sourced from M_P and the negative half of I_O is provided by M_N. Both ZCS and ZVS are realized in this case. With small dead-times for the on and off intervals of the power switches, shoot-through current can be eliminated, and high efficiency can be achieved.

However, when the load and/or coupling conditions change, the operating point will deviate from the ideal case (Case 1). Extra considerations are needed for Case 2 and Case 3, and are discussed as follows with the help of Fig. 7.6.

When $f_S < f_R$, the resonant load appears capacitive and I_O leads ahead of the voltage signal V_X. Therefore, I_{MP} goes negative across the zero point before M_P is switched off that defeats zero current switching (ZCS), which is originally scheduled for soft turn-off. Before M_N is turned on, I_O will keep flowing even M_P is already turned off. As a result, I_O finds its way through the parasitic diode D_P as indicated (by the shaded area) in the figure. Next, M_N is turned on with a non-zero current and a high V_{DS} (approximately equal to V_{DD}) that is essentially a hard turn-on. The parasitic capacitors at V_X are discharged that results in a large peak current and thus severe switching loss. In a similar fashion, in the next half cycle, M_N will undergo a soft turn-off and M_P a hard turn-on.

When $f_S > f_R$, the resonant load appears inductive and I_O lags behind V_X. After M_P is turned off with a non-zero current, and hence a hard turn-off, I_O will firstly discharge the parasitic capacitors at the node V_X, and will keep conducting through the parasitic diode D_N with $V_X = -V_D$, where V_D is the diode forward voltage drop. The conduction loss in this short dead-time duration is relatively large because of the large diode voltage drop. Next, after a small dead-time, M_N is turned on with a low V_{DS}. This can be considered as zero-voltage switching (ZVS) that resembles a soft turn-on with essentially no shoot-through current. Therefore, the switching loss is much reduced. In a similar fashion, in the next half cycle, M_N will undergo a hard turn-off and M_P a soft turn-on. As explained in the previous section, an MOS transistor has no problem in turning off its current, and thus ZVS is preferred.

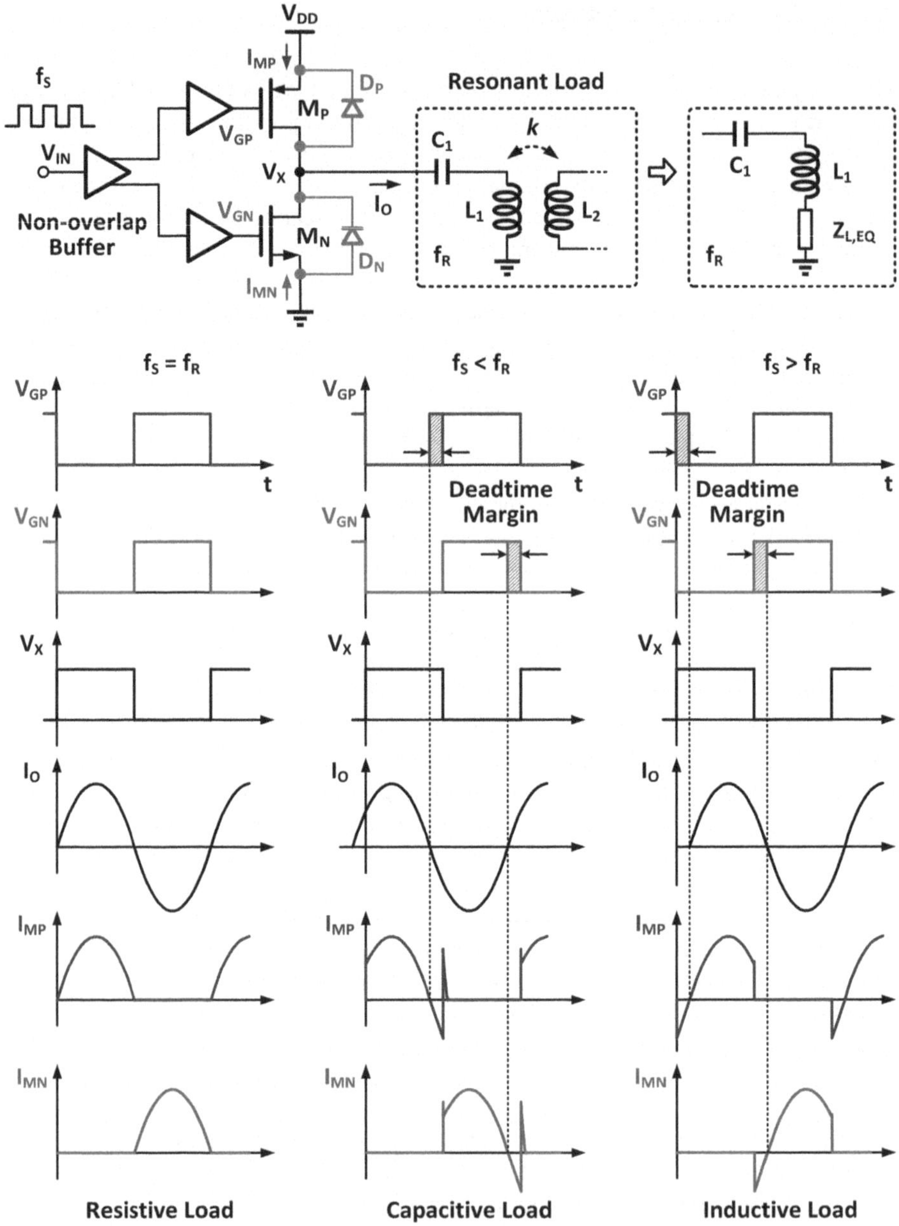

Fig. 7.5 The voltage and current waveforms of a Class-D PA in a resonant WPT system, with $f_S = f_R$ (resistive equivalent load); $f_S < f_R$ (capacitive equivalent load); $f_S > f_R$ (inductive equivalent load)

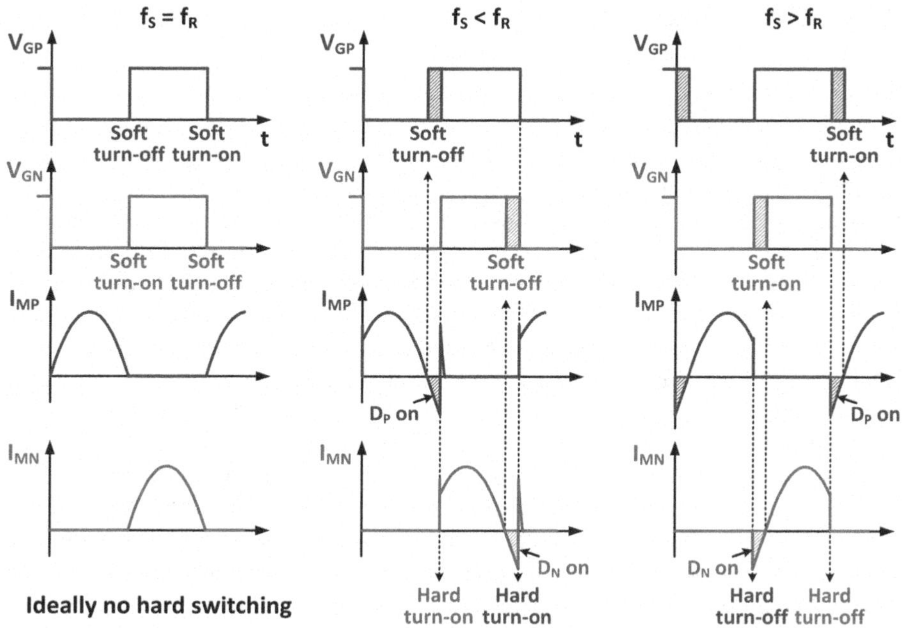

Fig. 7.6 The voltage and current waveforms of the power transistors in Class-D PA with soft switching considerations

The shaded areas in Fig. 7.6 represent the tolerances (acceptable dead-time margins) of the dead-time between the two power switches [22]. Setting the dead-time within the margin can guarantee a continuous I_O without staying at zero for a short duration. Conversely, the maximum allowed dead-time increases as f_S deviates more from f_R, because the time interval during which the switch current is negative becomes longer. When $f_S = f_R$, the shortest dead-time is required.

When the Class-D PA is driving an inductive load, larger parasitic capacitance is allowed at the node V_X, because the energy of the parasitic capacitors is simply transferred to the resonant tank without loss. A conventional Class-D PA generates electromagnetic interference (EMI) due to switching noise and high-frequency ringing at the switching nodes, but one may intentionally add a capacitor between V_X and ground to smooth out the switching at V_X [23]. In so doing, the V_X waveform is smoothen and contains a smaller amount of harmonics than the square wave of a conventional Class-D PA. Consequently, EMI and noise level are reduced. However, a longer switch dead-time is needed to guarantee that ZVS is realized, and the maximum output power will thus be reduced.

7.2.2 ZVS Class-D PA

As mentioned above, the conventional Class-D PA incurs high losses due to the parasitic capacitances at the V_X node, and therefore must be operating at $f_S > f_R$ with the load tuned to be slightly inductive. However, such an operating point of shifting away from the resonant frequency will increase the circulating energy between the coil and the PA, and consequently reduces the transmission efficiency.

Figure 7.7 shows a variation of the conventional Class-D PA with additional ZVS inductor L_{ZVS} and capacitor C_{ZVS} for the WPT system by EPC Corporation using enhancement-mode GaN devices [8, 9]. In this configuration, the ZVS tank circuit does not operate at resonance, but rather as a no-load buck converter. Let us refer to Fig. 7.7b. The peak of I_{ZVS} occurs at the ZVS point of $V_X = 0$, and it provides the necessary charging/discharging current for the V_X node. The value of L_{ZVS} depends on the supply voltage V_{DD}, parasitic capacitance at the V_X node, the slew rate at V_X and the immunity margin for shifts in the load impedance.

This scheme has two main drawbacks. The first is the extra cost brought by the bulky additional ZVS inductor and capacitor, making it not suitable for miniature applications. The second is that I_{ZVS} introduces extra conduction loss through the power transistors, limiting the light-load efficiency and the maximum output capability.

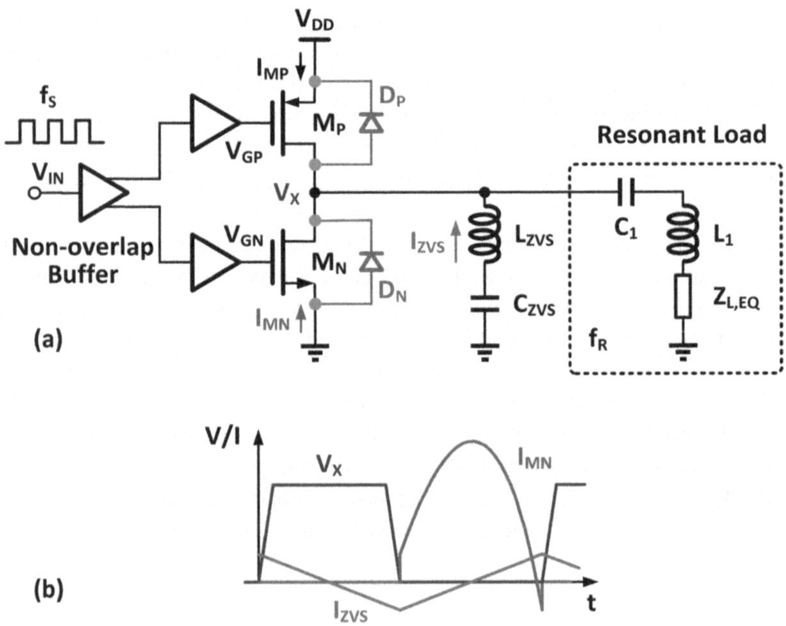

Fig. 7.7 (**a**) A ZVS variation to the conventional Class-D PA, and (**b**) its voltage and current waveforms

7.3 Class-E PA

The high-efficiency Class-E PA was invented by N. O. Sokal (the father) and A. D. Sokal (the son) in 1975 [12]. Figure 7.8a shows a simple diagram of the Class-E PA, and Fig. 7.8b shows the conceptual voltage and current waveforms operating in the steady-state. The Class-E PA of a WPT transmitter consists of a single switch M_N, an RF choke L_C, a shunt capacitor C_P, and an LC series resonant tank that contains C_S and the coupling coils L_1 and L_2. The inductance value of L_C is usually set to be sufficiently high such that its AC current is small compared to the DC current, but that is not a must for the design [24]. The parasitic junction capacitor of M_N is now part of C_P. The load network C_P, C_S and L_1 is a combined series-parallel resonant circuit, which is also known as a multi-frequency network.

The Class-E operation principle discussed in [12] covers a variety of circuit topologies that contain a switch and a load network, as long as they satisfy all three specific objectives for the collector voltage (drain voltage for MOSFET) and current waveforms: (1) the rise of the voltage across the transistor at turn-off should be delayed until after the transistor is off; (2) the drain voltage should go back to zero when the transistor starts to turn on; and (3) the slope of the drain voltage should be zero when the transistor starts to turn on. Therefore, the Class-E PA can achieve high power-conversion efficiency at high frequencies. In the past four decades, it finds various high-power and/or high-efficiency applications in

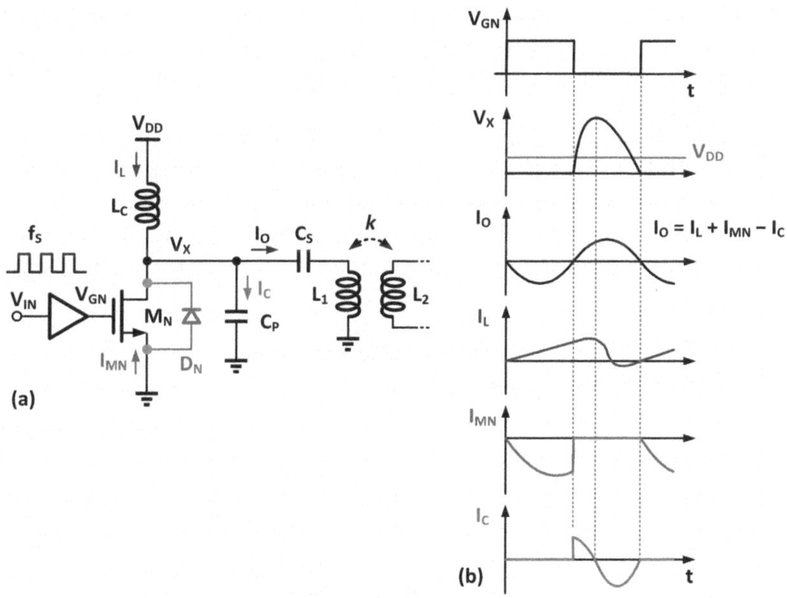

Fig. 7.8 (**a**) A simple diagram of Class-E PA in a resonant WPT system, and (**b**) the voltage and current waveforms in the steady-state

communication transceivers, electronic ballasts for fluorescent lighting, heat induction, electrosurgical generator and wireless power transfer systems [17, 25–27].

When the PA is driving a series-resonant tank, as in the conventional Class-D case, increasing the coil current by simply lowering the equivalent series resistances (ESRs) would require significant reduction of the switch on-resistance and the coil parasitic resistance. Therefore, large transistors should be used, which would increase the gate drive loss and add stray capacitance to the V_X node.

When the PA is driving a parallel-resonant tank, the reactances cancel each other at resonance. The large AC current required in the primary coil for large magnetic field is mainly provided by the resonant capacitor in parallel, which relaxes the current handling capability of the active device(s). However, the parallel-resonant tank induces high voltage stress to the driver device(s), and high-voltage devices such as LDMOS should be used, but such devices would occupy large area and may degrade the high-frequency performance.

To deal with the above cases, a compromised solution is to use a combined series-parallel resonant tank, which has been employed in the Class-E configuration. In this case, the active device deals with both low current and low voltage while delivering a large coil current. The impedance curve of the multi-frequency load network is plotted in Fig. 7.9. The lower peak of the frequency response is caused by the series resonance of L_1 and C_S. As the frequency increases, the reactance of the series-resonant tank increases, and with C_P, forms the parallel resonance frequency peak. The Class-E point is just located between the series resonance and parallel resonance frequency peaks, at where the power loss of the driver is minimized [1]. In this case, the voltage and current waveforms are 180° out of phase. For a set of component values of C_P, C_S and L_1, the Class-E frequency point depends on the Q of the load network, while the minimum-loss operation occurs at a critical Q.

The design parameters of a Class-E PA are correlated, and therefore a priori determination of the exact Class-E point (the critical Q) is difficult [1]. When the Q of the load network is too high, the ringing of the switching voltage V_X is underdamped, which would cause V_X to swing well below ground and turns on the parasitic diode of M_N, causing more conduction loss. If the Q is too low, the ringing

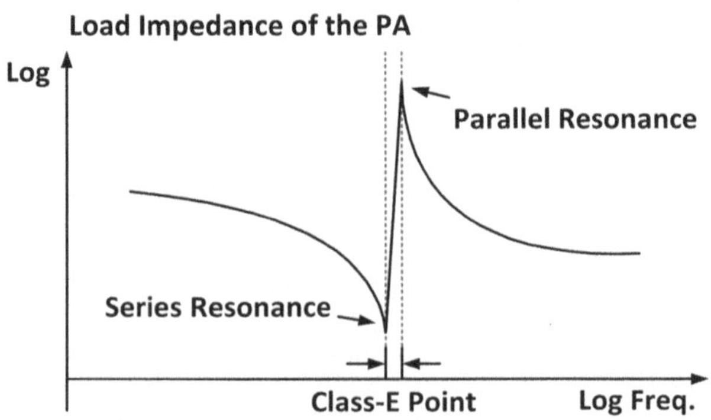

Fig. 7.9 Impedance of the load network of the Class-E PA

of V_X is over-damped, and V_X may not return to zero voltage fast enough and results in unfavorable non-zero voltage switching that introduces extra switching loss.

The above two cases can be referred as sub-optimum Class-E operation [14], and a critical Q is required for the minimum-loss operation. However, the required condition is sensitive to frequency and component variations, and may require a closed-loop tracking compensation scheme [1, 2]. Effects of parameter variations, including load impedance variation, shunt reactance variation, frequency variation and the duty cycle variation were discussed in [24, 28]. It is also suggested that modulating the Class-E PA output power with pulse-width modulation (PWM) would degrade the efficiency but would not be excessive. In addition, the load resistance and the supply voltage V_{DD} are related by the requirement of delivering a specified output power to the load from V_{DD} [12]. Therefore, the output power can be tuned by modulating the supply voltage of the Class-E PA.

7.4 Summary and Discussion

This chapter discussed two popular types of DC-AC converters, known as power amplifiers in the context of WPT transmitters, which are the Class-D and Class-E amplifiers. Some essential knowledge on integration processes, power loss components, as well as soft switching mechanisms are introduced.

Both Class-D and Class-E PAs have an ideal efficiency of 100%, but in practice, Class-E PAs can be more efficient than Class-D PAs. Class-D PAs suffer from the possibility of turning on and off both high-side and low-side transistors simultaneously, leading to efficiency loss at high frequencies; and the gate-drive circuits have to have long enough dead-times. Long dead-times run the risk of turning on the parasitic diodes that introduce additional conduction loss. A CMOS Class-D PA should operate at $f_S > f_R$, that is, it should drive an inductive load to achieve zero-voltage switching that reduces switching loss and EMI. To achieve high efficiency, a Class-D PA should work with a duty ratio of 50%, while this is not necessary for a Class-E PA. The Class-E PA can be designed to drive a combined series-parallel resonant load network. The single power switch only needs to provide part of the total resonant current and thus results in lower conduction loss. Moreover, the Class-E switching condition ensures zero-voltage switching at zero voltage-slope and reduces the switching loss to zero during the turn-on transition.

The output power capability of Class-D and Class-E PAs was analyzed in [14], and the calculation method has been used in [29] for the Class-DE PA. Their output power capabilities P_{MAX} have been normalized to the product of the peak current stress I_P and the peak voltage stress V_P imposed on the switches as given below:

$$P_{MAX} = \frac{P_{OUT}}{V_P I_P} \tag{7.1}$$

For Class-D and Class DE PAs, the peak voltage is V_{DD} only, while the peak voltage for Class-E PAs is $3.56V_{DD}$ [14]. Therefore, the normalized output power capability of the Class-D PA is higher than that of the Class-E PA, which may limit the application of the Class-E PA in high-power applications such as fast wireless charging systems.

Employing the Class-E PA for WPT requires detailed knowledge of the coupling coils and how the load would affect the impedance seen by the PA. The design is further complicated by coupling-variation between the coils and power-variation due to the changing load. A dedicated closed-loop control for the Class-E PA is needed to achieve high efficiency over component variations and a wide range of conditions. Therefore, the Class-D PA may still be a more robust selection for WPT systems.

References

1. Troyk PR, Schwan MAK (1992) Closed-loop class E transcutaneous power and data link for microImplants. IEEE Trans Biomed Eng 39:589–599. doi:10.1109/10.141197
2. Kendir G, Liu W, Wang G et al (2005) An optimal design methodology for inductive power link with class-E amplifier. IEEE Trans Circuits Syst I Regul Pap 52:857–866. doi:10.1109/TCSI.2005.846208
3. Tomita K, Shinoda R, Kuroda T, Ishikuro H (2012) 1-W 3.3–16.3-V boosting wireless power transfer circuits with vector summing power controller. IEEE J Solid State Circuits 47:2576–2585. doi:10.1109/JSSC.2012.2211698
4. Choi J-H, Yeo S-K, Park S, Lee J-S, Cho G-H (2013) Resonant regulating rectifiers (3R) operating for 6.78 MHz resonant wireless power transfer (RWPT). IEEE J Solid State Circuits 48:2989–3001. doi:10.1109/JSSC.2013.2287592
5. Li X, Lu Y, Tsui C-Y, Ki W-H (2014) An adaptive wireless powering and data telemetry system for optic nerve stimulation. In: 2014 IEEE international symposium on circuits and systems (ISCAS), pp 1404–1407. doi:10.1109/ISCAS.2014.6865407
6. Li X, Meng X, Tsui C-Y, Ki W-H (2016) Reconfigurable resonant regulating rectifier with primary equalization for extended coupling- and loading-range in bio-implant wireless power transfer. IEEE Trans Biomed Circuits Syst:1–1. doi:10.1109/TBCAS.2015.2503418
7. Kosuge A, Hashiba J, Kawajiri T et al (2016) An inductively powered wireless solid-state drive system with merged error correction of high-speed wireless data links and NAND flash memories. IEEE J Solid State Circuits 51:1041–1050. doi:10.1109/JSSC.2015.2512959
8. Lidow A, de Rooij M (2014, May) Performance evaluation of enhancement-mode GaN transistors in Class-D and Class-E wireless power transfer systems. Bodo Magaz: 56–60
9. de Rooij MA (2015) The ZVS voltage-mode class-D amplifier, an eGaN® FET-enabled topology for highly resonant wireless energy transfer. In: 2015 IEEE applied power electronics conference and exposition (APEC), pp 1608–1613
10. Rose M, Bergveld HJ (2016) Integration trends in monolithic power ICs: application and technology challenges. IEEE J Solid-State Circuits:1–10. doi:10.1109/JSSC.2016.2566612
11. El-Hamamsy S-A (1994) Design of high-efficiency RF Class-D power amplifier. IEEE Trans Power Electron 9:297–308. doi:10.1109/63.311263
12. Sokal NO, Sokal A (1975) Class E-A new class of high-efficiency tuned single-ended switching power amplifiers. IEEE J Solid State Circuits 10:168–176. doi:10.1109/JSSC.1975.1050582

13. Sokal NO (1981) Class E high-efficiency switching-mode tuned power amplifier with only one inductor and one capacitor in load network-approximate analysis. IEEE J Solid State Circuits 16:380–384. doi:10.1109/JSSC.1981.1051605
14. Raab FH (1977) Idealized operation of the class E tuned power amplifier. IEEE Trans Circuits Syst 24:725–735. doi:10.1109/TCS.1977.1084296
15. Redl R, Molnar B, Sokal NO (1986) Class E resonant regulated DC/DC power converters: analysis of operations, and experimental results at 1.5 MHz. IEEE Trans Power Electron PE 1:111–120. doi:10.1109/TPEL.1986.4766289
16. Liu KH, Lee FCY (1990) Zero-voltage switching technique in DC/DC converters. IEEE Trans Power Electron 5:293–304. doi:10.1109/63.56520
17. Erickson RW, Dragan M (2001) Fundamentals of power electronics, 2nd edn. Norwell, Kluwer
18. Baxandall PJ (1959) Transistor sine-wave LC oscillators. Some general considerations and new developments. Proc IEE Part B Electron Commun Eng 106:748–758. doi:10.1049/pi-b-2.1959.0141
19. Berkhout M (2003) An integrated 200-W class-D audio amplifier. IEEE J Solid State Circuits 38:1198–1206. doi:10.1109/JSSC.2003.813238
20. Ma H, van der Zee R, Nauta B (2015) A high-voltage Class-D power amplifier with switching frequency regulation for improved high-efficiency output power range. IEEE J Solid State Circuits 50:1451–1462. doi:10.1109/JSSC.2015.2421994
21. Amir S, van der Zee R, Nauta B (2015) An improved modeling and analysis technique for peak current-mode control-based boost converters. IEEE Trans Power Electron 30:5309–5317. doi:10.1109/TPEL.2014.2368176
22. Kazimierczuk MK (1991) Class D voltage-switching MOSFET power amplifier. IEE Proc B Electr Power Appl 138:285–296. doi:10.1049/ip-b.1991.0035
23. Kazimierczuk MK, Szaraniev W (1992) Class-D zero-voltage-switching inverter with only one shunt capacitor. IEE Proc B Elect Power Appl 139:449–456. doi:10.1049/ip-b.1992.0055
24. Kazimierczuk M, Puczko K (1987) Exact analysis of class E tuned power amplifier at any Q and switch duty cycle. IEEE Trans Circuits Syst 34:149–159. doi:10.1109/TCS.1987.1086114
25. Tsai K-C, Gray PR (1999) A 1.9-GHz, 1-W CMOS class-E power amplifier for wireless communications. IEEE J Solid State Circuits 34:962–970. doi:10.1109/4.772411
26. Wang G, Liu W, Sivaprakasam M, Kendir G (2005) Design and analysis of an adaptive transcutaneous power telemetry for biomedical implants. IEEE Trans Circuits Syst I Regul Pap 52:2109–2117. doi:10.1109/TCSI.2005.852923
27. Low ZN, Chinga R, Tseng R, Lin J (2009) Design and test of a high-power high-efficiency loosely coupled planar wireless power transfer system. IEEE Trans Ind Electron 56:1801–1812. doi:10.1109/TIE.2008.2010110
28. Raab FH (1978) Effects of circuit variations on the class E tuned power amplifier. IEEE J Solid State Circuits 13:239–247. doi:10.1109/JSSC.1978.1051026
29. Koizumi H, Suetsugu T, Fujii M et al (1996) Class DE high-efficiency tuned power amplifier. IEEE Trans Circuits Syst I Fundam Theory Appl 43:51–60. doi:10.1109/81.481461

Chapter 8
Conclusions and Future Works

Abstract In this chapter, we summarize the contents of the book and make a few remarks on the design perspectives of each building block of the wireless power transfer system. Potential research and development directions are suggested for the consideration of research students and engineers who are working on or going to work on this promising area.

Keywords Wireless power transfer • Inductive coupling • Wireless charging • DC-DC converter • Voltage regulator • Power management IC

8.1 Concluding Remarks of the Book

Wireless power transfer (WPT) for a broad range of applications is projected to have an exponential growth in demand, with an enormous number of new devices and products to be enabled by this powerful technology. A WPT system involves almost all kinds of power converters such as AC-DC, DC-AC and DC-DC converters, and a circuit designer should have deep understandings of the fundamentals and operation principles of each type of power converters. To design an efficient wireless power transfer system, an engineer needs to know both power processing and signal processing, and also coil design. These requirements impose technical challenges to both the circuit and the system designers.

In this book, we started in Chap. 1 with the motivations and background of wireless power transfer. Then, in Chap. 2, we reviewed the system architectures for WPT receivers, and focused on output voltage regulation and passive component reduction. We also analyzed the characteristics of inductive coil coupling in Chap. 3. Subsequently, we discussed the CMOS integrated-circuit design that focused on low-power wireless power transfer systems in Chaps. 4, 5, 6 and 7. Design examples were studied in details in individual chapters, and are summarized as follows.

In Chap. 2, a three-level single-inductor dual-output DC-DC converter in cooperation with an active voltage doubler was designed for a wireless power receiver of which the load needs multiple step-up/−down supply rails. Small output voltage ripples were obtained at all output voltages, while small size passive components

Y. Lu, W.-H. Ki, *CMOS Integrated Circuit Design for Wireless Power Transfer*, Analog Circuits and Signal Processing, DOI 10.1007/978-981-10-2615-7_8

could be used due to high switching frequency and discontinuous-current mode operation.

A couple of active rectifiers for near-field WPT receivers were designed and analyzed in Chap. 4. Reverse current control schemes were reviewed and summarized, while a competent and advantageous scheme with additional analog auto-adjust loops for active rectifiers was introduced. Delay-locked-loop (DLL) based active rectifiers, which were found to be more suitable for the series-resonant receiving tank and/or high WPT frequency cases, were studied as well.

Chapters 5 and 6 introduced linear regulator and switching converter designs, respectively. Two fully-integrated low-dropout regulators with full-spectrum power supply ripple rejection were designed for suppressing high-frequency supply ripples at the WPT frequency and its harmonics. Circuit and architectural techniques for fast transient response and small output ripples were discussed as well.

Chapter 7 compared the Class-D and Class-E power amplifiers (PAs) for wireless power transmitters. It is suggested that the knowledge on power electronics and especially on resonant converters can be very helpful for optimizing the PA efficiency.

8.2 Suggested Future Works

The following research directions are of interest to the authors of this book.

Device-to-device wireless power transfer is an emerging direction for consumer electronics and internet-of-things, and is creating a new eco-system for wireless charging [1]. For wireless power transfer among devices (for example, mobile phones), both source energy and output power level are limited, and therefore, the battery-to-battery power transfer efficiency is one of the key specifications.

Ultrasound WPT could be a good alternative to electromagnetic (EM) wave WPT for implantable medical devices [2]. It is known that the propagation loss in water of acoustic waves is much lower than that of the EM waves, and it is relatively safe for the tissue to absorb ultrasound energy at higher power density [3]. Meanwhile, the relatively low operation frequency compared to the EM waves would result in higher conversion efficiency for the power converters.

Wireless charging technology using the Wi-Fi frequency bands (2.45 or 5.8 GHz) for longer WPT distance (up to the 10-m range) and for better spatial freedom [4] is attractive especially for in-office and in-car power transmission. To increase the received power, a complex antenna phased array could be designed to precisely send power to the device-under-charging with multiple energy beams.

We hope the readers enjoy reading this book, and continue or start their research in this interesting and promising area.

References

1. Huang M, Lu Y, et al (2017) A reconfigurable bidirectional wireless power transceiver with maximum current charging mode and 58.6% battery-to-battery efficiency. In: 2017 IEEE international solid- state circuits conference – (ISSCC), pp 376–377. doi:10.1109/ISSCC.2017.7870418
2. Meng M, Kiani M (2016) Design and optimization of ultrasonic wireless power transmission links for millimeter-sized biomedical implants. IEEE Transac Biomed Circuits Syst:1–10. doi:10.1109/TBCAS.2016.2583783
3. van Schuylenbergh K, Puers R (2009) Inductive powering. Springer, Dordrecht
4. Branscombe M (2013) Wireless power: could Cota make it long-distance and mainstream? http://www.qiwireless.com/wireless-power-cota-make-long-distance-mainstream. Accessed 19 Oct 2016

The manufacturer's authorised representative in the EU is Springer Nature Customer Service Centre GmbH, Europaplatz 3, 69115 Heidelberg, Germany. If you have any concerns regarding our products, please contact ProductSafety@springernature.com

Printed and bound by CPI Group (UK) Ltd, Croydon, CR0 4YY
25/09/2026
02222869-0019